Options for Minimizing Fuel-Related Contaminants in Lakes and Reservoirs

The mission of the Awwa Research Foundation (AwwaRF) is to advance the science of water to improve the quality of life. Funded primarily through annual subscription payments from over 1,000 utilities, consulting firms, and manufacturers in North America and abroad, AwwaRF sponsors research on all aspects of drinking water, including supply and resources, treatment, monitoring and analysis, distribution, management, and health effects.

From its headquarters in Denver, Colorado, the AwwaRF staff directs and supports the efforts of over 700 volunteers, who are the heart of the research program. These volunteers, serving on various boards and committees, use their expertise to select and monitor research studies to benefit the entire drinking water community.

Research findings are disseminated through a number of technology transfer activities, including research reports, conferences, videotape summaries, and periodicals.

Options for Minimizing Fuel-Related Contaminants in Lakes and Reservoirs

Prepared by:
University of California, Davis, CA
and
CH2M HILL, Oakland, CA
and
Contra Costa Water District, Concord, CA

Jointly sponsored by:
Awwa Research Foundation
6666 West Quincy Avenue, Denver, CO 80235-3098
and
California Department of Health Services
Sacramento, CA
and
Contra Costa Water District
Concord, CA

Published by:

Awwa Research Foundation
Advancing the Science of Water®

IWA Publishing

DISCLAIMER

Published by IWA Publishing, Alliance House, 12 Caxton Street, London SW1H 0QS, UK

Telephone: +44 (0) 20 7654 5500; Fax: +44 (0) 20 7654 5555; Email: publications@iwap.co.uk;
Web: **www.iwapublishing.com**

AwwaRF report number 90955F. Originally published by AwwaRF for its subscribers in 2003.
IWA Publishing version published 2004.

ISBN 1 84339 8648

CONTENTS

LIST OF TABLES ... ix
LIST OF FIGURES ... xi
FOREWORD ... xiii
ACKNOWLEDGMENTS ... xv
EXECUTIVE SUMMARY ... xvii
CHAPTER 1: INTRODUCTION, BACKGROUND, AND PROJECT OBJECTIVES ... 1
INTRODUCTION AND BACKGROUND ... 1
History of MTBE and BTEX in the U.S. ... 2
Current Concerns ... 3
MTBE/BTEX Sources, Transport, and Exposure Pathways ... 5
Regulatory Setting ... 10
REVIEW OF PREVIOUS STUDIES ON MTBE ... 15
PROJECT OBJECTIVES ... 17
Project Significance and Applicability ... 18
RESEARCH APPROACH ... 19
CHAPTER 2: MODELING MTBE AND BTEX ... 21
INTRODUCTION ... 21
DYNAMIC LAKE MODEL (DLM) ... 22
Model Structure ... 22
Model Processes ... 23
Model Inputs ... 24
Model Outputs ... 24
DESCRIPTION OF DLM-1998 ... 24
Simulations ... 25
MTBE Volatilization Sub-model ... 25
Inputs and Outputs ... 25
Results ... 25
Model Review ... 26
MODIFICATIONS TO DLM-1998 ... 26

Introduction......26
Model Structure......27
Model Input Data......28
CHAPTER 3: RESEARCH FINDINGS......31
RESERVOIR MANAGEMENT QUESTIONNAIRE......31
Questionnaire Description and Objectives......32
Results......33
Marine Engine Technologies......40
Operating Principles......41
Review of Existing Emission Studies......43
DONNER LAKE TEST SITE......63
Physical Characteristics of the Lake......65
Year 2000 Boating Survey at Donner Lake (Census)......71
Census Data......74
MTBE/BTEX Water Quality Monitoring Program......83
CHAPTER 4: MODEL CALIBRATION AND VERIFICATION......95
INTRODUCTION......95
PRELIMINARY CONSIDERATIONS......95
Quality of Input Data......95
Analysis of Boating Census Data......98
DLM-2002 CALIBRATION......102
Hydrodynamic Calibration......102
VOC Calibration......103
MODEL VERIFICATION......107
Calero Reservoir......108
Lake Perris......109
COMPARISON OF DLM-1998 AND DLM-2002 MODELS......109
COMPARISON OF 1998 AND 2000 BOATING SEASONS......111
Number of Boats......111
Weather......112
CHAPTER 5: SPREADSHEET MODEL......115

INTRODUCTION 115
METHOD OF ESTIMATION 115
TESTING 116
ESTIMATOR PORTABILITY 121
INSTRUCTIONS FOR USE 122
DISCLAIMERS 122
CONCLUSIONS 123
CHAPTER 6: RESERVOIR MANAGEMENT OPTIONS 125
INTRODUCTION 125
Description of Methodology and Assumptions 125
DESCRIPTION OF MANAGEMENT OPTIONS 126
Source Control through Management of Recreational Boating 126
Treatment 133
Operations Management 140
Economic Considerations 142
APPENDIX A: LIST OF VOCS 149
APPENDIX B: DETAILED DESCRIPTION OF MODEL 153
APPENDIX C: LIST OF AGENCIES CONTACTED 161
APPENDIX D: MTBE AND OTHER FUEL-RELATED CONTAMINANTS QUESTIONNAIRE 2001 171
APPENDIX E: SUMMARY OF SURVEY FINDINGS 177
APPENDIX F: MARINE ENGINE TECHNOLOGIES 199
APPENDIX G: BOAT RAMP DATABASE EXAMPLE 203
APPENDIX H: SAMPLING PROGRAM DESIGN/RESULTS 205
APPENDIX I: USER GUIDE TO SPREADSHEET MODEL 223
APPENDIX J: EVALUATION OF THE ECONOMIC IMPACTS OF RECREATIONAL BOATING 227
APPENDIX K: COST SUMMARY FOR TREATMENT TECHNOLOGIES 235
REFERENCES 241
ABBREVIATIONS 251

TABLES

1.1	Physical and chemical properties of BTEX compounds and MTBE	7
1.2	State agencies with MTBE drinking water standards	13
3.1	Engine emission data currently available – February 2001	44
3.2	Summary of the results of MTBE, BTEX, and benzene emission data from CARB (2001) and Fiore et al. (1998)	53
3.3	Emission fractions of MTBE from marine engines into lake water during use	58
3.4	Average meteorological values for July and August	67
3.5	Donner Lake temperature extremes during study period	68
3.6	Wind extremes	70
3.7	Engine types	73
3.8	Monthly boat use at Donner Lake during the 2000 boating season	76
3.9	Activity fractions for 2000 boating season at Donner Lake	77
3.10	List of marine engines and codes	79
3.11	Calculated volume (gallons) of VOCs entering Donner Lake from marine engine emissions	83
3.12	Reporting levels or analytical limits of detection at Donner Lake	86
4.1	Engine emission rates	97
4.2	Fractions of VOC constituents in fuel	98
4.3	Average daily level of boating activity, Donner Lake 2000	99
4.4	Percentage of engine types, Donner Lake 2000	100
4.5	Average activity duration, Donner Lake 2000	100
4.6	Fuel consumption rates, Donner Lake 2000	101
4.7	Average meteorological values for July and August	112
6.1	Cost considerations of options to manage MTBE in lakes and reservoirs	145

FIGURES

1.1 MTBE movement in the environment ... 9

3.1 MTBE monitoring data for three California reservoirs ... 37

3.2 Effect of boat use on seasonal variation of MTBE concentration in Donner Lake ... 48

3.3 Summary data for the median concentration of MTBE in those lakes and reservoirs that allowed and excluded motorized watercraft ... 50

3.4 Fuel use by engine ... 56

3.5 Estimated MTBE load by engine type ... 56

3.6 Effect of the two-stroke ban at Lake Tahoe ... 58

3.7 Census and water quality monitoring program locations at Donner Lake ... 64

3.8 Daily average air temperature, Donner 2000 ... 69

3.9 Daily average wind velocity, Donner 2000 ... 70

3.10 Estimated daily boat use and measured MTBE at Donner Lake ... 75

3.11 Average weekend boating activity July and August 2000 ... 78

3.12 Average weekday boating activity July and August 2000 ... 79

3.13 Estimated boat usage on Donner Lake by engine type during the 2000 boating season ... 81

3.14 Estimated engine fuel consumption by engine type for all motorized watercraft at Donner Lake during the 2000 boating season ... 82

3.15 Depth distribution of MTBE in Donner Lake ... 88

3.16 Seasonal distribution of epilimnetic MTBE in Donner Lake, May 22 – October 23, 2000 ... 89

3.17 Change in MTBE concentration at station 4 over a 10-day period in mid-August ... 90

3.18 Daily changes in MTBE in Donner Lake as measured on a mid-August weekend ... 92

4.1 Modeled temperature vs. measured temperature for wind multiplier = 1.2 ... 103

4.2 MTBE calibration runs ... 105

4.3 Calibration model runs for toluene ... 105

4.4 Calibration runs for benzene and ethylbenzene 106
4.5 Calibration results for xylene 107
4.6 Measured vs. Modeled MTBE in Calero Reservoir, 1997 108
4.7 Measured vs. Modeled MTBE in Lake Perris, 1997 109
4.8 Comparison of DLM-2002 and DLM-1998 using 1997 data 110
4.9 Comparison of DLM-2002 and DLM-1998 models 111
5.1 Spreadsheet estimator vs. DLM for MTBE in Donner Lake 117
5.2 Spreadsheet estimator vs. DLM for Toluene in Donner Lake 117
5.3 Spreadsheet estimator vs. DLM for Benzene in Donner Lake 118
5.4 Spreadsheet estimator vs. DLM for Ethylbenzene in Donner Lake 118
5.5 Spreadsheet estimator vs. DLM for Xylene in Donner Lake 119
5.6 Estimator vs. DLM for MTBE in Lake Perris, 1997 120
5.7 Estimator vs. DLM for MTBE in Calero Reservoir, 1997 120
5.8 Effect of turbidity on MTBE concentration 122
6.1 MTBE decay for three scenarios 127
6.2 Effects of weekend-only boating 128
6.3 Effect of engine mix on MTBE concentrations, medium density boating 129
6.4 Effect of engine mix on MTBE concentrations, high density boating 130
6.5 Effect of boating density on MTBE level, Donner Lake 130
6.6 Peak weekend effects after low density boating 131
6.7 Peak weekend effects after medium density boating 132
6.8 Peak weekend effects after high-density boating 132
6.9 Effect of MTBE content in fuel 134

FOREWORD

This publication is the result of a tailored collaborative study jointly sponsored by the Awwa Research Foundation (AwwaRF) and the State of California Department of Health Services (DHS). The "day-to-day" management of the study was conducted by the AwwaRF staff under an agreement between the AwwaRF and DHS.

The AwwaRF is a non-profit corporation dedicated to the implementation of a research effort to help water utilities respond to regulatory requirements, assist water utilities in providing the highest possible water quality, economically, and reliably, and respond to traditional high-priority concerns of the water industry. The AwwaRF research agenda is developed through a process of consultation with subscribers and drinking water professionals. A broad spectrum of water supply issues is addressed by the AwwaRF research agenda including resources, treatment and operations, distribution and storage, water quality and analysis, toxicology, economics, and management.

Under the umbrella of the AwwaRF Strategic Research Plan, the Research Advisory Council prioritizes the suggested projects based upon current and future needs, applicability, and past work. The recommendations are forwarded to the Awwa Board of Trustees for final selection. The funding of the research efforts is financially supported primarily from the AwwaRF Subscription Program whereby water utilities, consultants, and manufacturers make annual payments to subscribe to the AwwaRF research program.

AwwaRF also sponsors research projects through an unsolicited proposal process including the Collaborative Research, Research Application, and Tailored Collaboration Programs. The AwwaRF also participates in other joint research efforts with organizations such as the U.S. Environmental Protection Agency, the U.S. Bureau of Reclamation, and Association of California Water Agencies.

The State of California Department of Health Services (DHS) is authorized by state law to conduct research for the purpose of developing cost effective water treatment technologies to remove gasoline oxygenates and the development of strategies to protect drinking water supplies from gasoline oxygenate contamination. DHS implements the research program through cooperative agreements with federal, state, and local agencies, as well as academic institutions and in public water systems. This research program is financially supported by the DHS Drinking Water Treatment and Research Fund (DWTRF).

The selection of research projects is based on the recommendations of the DWTRF Research Advisory Committee (RAC), which is comprised of representatives from public water systems, oil companies and petroleum product distributors, environmental and consumer groups, and the Department of Health Services.

This collaborative research project was undertaken to provide practical advice to water utilities considering the benefits of using lakes and/or reservoirs for both recreational and drinking water purposes, without introducing unacceptable health risks into the water resource from Methyl Tertiary Butyl Ether (MTBE) or other fuel related contaminants. The guidance provided in this document is applicable to all water utilities and water management agencies with large surface water storage facilities that currently allow recreational boating or are considering their use on the surface water storage facilities in the future.

Edmund G. Archuleta, P.E.
Chair, Board of Trustees
Awwa Research Foundation

James F. Manwaring, P.E.
Executive Director
Awwa Research Foundation

Gary D. Hoffman, P.E.
Chair, Research Advisory Committee
California Department of Health Services

David P. Spath, Ph.D., P.E., Chief
Division of Drinking Water &
Environmental Management
California Department of Heath Services

ACKNOWLEDGMENTS

The primary authors of this report are:

Brant Allen, Staff Research Associate, University of California, Davis

Scott Heald, Research Assistant, University of California, Davis

John Reuter, Associate Research Professor, University of California, Davis

Geoff Schladow, Professor of Water Resources and Environmental Engineering, University of California, Davis

Radha Gopalan, Staff Engineer, CH2MHILL

Diana Sokolove, Associate Planner, CH2MHILL

Alice Towey, Staff Engineer, CH2MHILL

The authors of this report are indebted to the following water utilities and individuals for their co-operation and participation in this project:

Bevin Beaudet, Vice President, CH2MHILL

David Bender, Hydrologist, U.S. Geological Survey (Project Advisory Committee [PAC] member)

Elizabeth Cooper, Intern, University of California, Davis

John Gaston, Vice President, CH2MHILL

Tracy Hemmeter, Program Administrator with Groundwater Management Unit, Santa Clara Valley Water District (PAC member)

Alan Highstreet, Senior Economist, CH2MHILL

David Huey, Water Operations Manager, Contra Costa Water District

Gary Hoffman, Chair, Research Advisory Committee, Department of Health Services

Dale Newkirk, Contra Costa Water District, Retired

Richard Talley, Water Quality Manager, Fort Worth Water Department

EXECUTIVE SUMMARY

This manual, commissioned by the DHS and Awwa Research Foundation (AwwaRF), provides information for water utilities on controlling MTBE and BTEX levels in drinking water reservoirs. These components of modern fuel typically enter reservoirs via recreational boating. A model has been developed to estimate the fate and transport of these constituents based on features such as reservoir size, boating density, and weather conditions. A simplified version of this model is presented in spreadsheet form to quickly and simply help reservoir managers predict whether MTBE/BTEX levels in their particular reservoir deserve more detailed investigation. Options for controlling MTBE and BTEX levels, and the corresponding economic effects, are also presented.

BACKGROUND

Oxygenates have been added to fuel since the late 1970s in order to improve the octane of motor fuel. Since then, several federal regulations have come into play requiring their inclusion to reduce air emissions. MTBE has become one of the most common fuel additives in American gasoline.

Currently MTBE and BTEX constituents (Benzene, Toluene, Ethylbenzene, and Xylenes) contribute largely to the make-up of commercial gasoline. Over the past ten years, concern has risen over the presence of these chemicals in freshwater sources such as lakes and reservoirs. While they enter the water via precipitation and dry fallout, these products are introduced primarily by gas powered aquatic recreational vehicles. In particular, two-stroke engines, such as jet skis, can release as much as half of their fuel directly into the water, unburned.

The realization that MTBE and BTEX constituents were entering the water supply caused concern for many water utilities due to the potential health and aesthetic effects associated with these substances. BTEX chemicals are known to pose a threat to human health; Benzene, for example, is a recognized human carcinogen, and Toluene is a developmental toxicant. MTBE has been identified as a potential human carcinogen and the State of California has established both primary and secondary drinking water standards to address the health risks and aesthetic impacts to drinking water

PROJECT OBJECTIVES

There are many possible methods for controlling the levels of MTBE and BTEX in water supplies from reservoirs. Thus far, none of these options have been thoroughly studied to quantify their effectiveness in reducing MTBE and BTEX and their economic consequences. The purpose of this report is to provide such guidance to reservoir managers who are weighing the various alternatives.

One goal of this research was to evaluate ways in which the concentrations of MTBE and BTEX could be controlled while still allowing recreational boating on reservoirs. This work culminated in the production of a new version of the University of California at Davis Dynamic Lake Model (DLM) for the purpose of quantifying the fate and transport of these contaminants under varying conditions. The new version of DLM will be referred to in this manual as DLM-2002. To complement this detailed model, a simple-to-use spreadsheet model (estimator) has been developed. Directions for downloading this estimator are included with this manual. The estimator allows reservoir managers to quickly estimate the impacts of different control strategies on their particular reservoirs.

MODELING MTBE AND BTEX

Most of the previous work on modeling MTBE and BTEX in surface waters focused on representing volatilization. However, this type of model is not always appropriate in deep lakes and reservoirs, where density stratification and lower turbulence intensities prevail. Therefore, a new type of model was needed.

DLM, first used by the University of California at Davis (UCD) in 1998 for the investigation of MTBE in reservoirs, is a process-based, one-dimensional hydrodynamic model. This model assumes that lateral variations are insignificant compared to vertical variation. It represents a reservoir's vertical profile as a set of Lagrangian layers that can move up or down and contract or expand in response to inflows and outflows, and to mixing events.

The original 1998 model, known as DLM-1998,[1] was modified for this report with the addition of more recent data on MTBE and field studies. The model was also adapted to monitor BTEX fate and transport. The accuracy of the original 1998 version was limited due to a lack of sufficiently detailed input data such as boat use and emission rates. This project sought to include a better accounting of boat traffic and improved data on engine emissions.

Input files for the current model include meteorological data, inflow, outflow, physical constants, reservoir volume, initial profile, and recreational boat use data. The resulting output files include daily profiles of MTBE, temperature, and salinity.

RESEARCH METHODS AND FINDINGS

Review of Existing Emissions Studies

Much of the original research on surface water contamination due to motorized watercraft was conducted in California. This includes the year 2000 survey of southern California reservoirs conducted by Dale et al. (Dale 2000), work done by UCD on Donner Lake and Lake Tahoe in the late 1990s, and a UC survey of California surface drinking water supplies conducted in 1998 (Reuter et al. 1998). Based on these and other studies, it became clear to researchers that marine engines are largely to blame for the intrusion of MTBE into surface water supplies. The next step was to determine which type of engine was the largest contributor. Conventional two-stroke engines ranked worst in terms of the amount of fuel emitted; both the fuel port and the exhaust port are open at the same time, allowing raw oil and gas to escape with the exhaust gases. Several improvements have been made to conventional two-stroke engines to improve fuel efficiency and decrease emissions. Electric engines and four-stroke engines are cleaner options.

In 1998, a team of researchers including UCD, the University of Nevada at Reno (UNR), the United States Geological Survey (USGS) and the California State Water Resources Control Board (SWRCB) mounted an investigation of fuel pollution entering Lake Tahoe from various watercraft. Their work ultimately found that two-stroke engines emitted considerably more

[1] In this manual, "DLM-1998" and "DLM-2002" refer to specific versions of the UC Davis Dynamic Lake Model. "DLM" (without a version date) refers to the model in general, including both versions.

gasoline than four-stroke engines did. The results of these studies were used to provide input data for DLM-1998, including emission rates and emission ratios.

Reservoir Management Questionnaire

A survey of water agencies operating lakes and reservoirs was conducted to collect information on currently practiced options for managing MTBE in lakes and reservoirs. Control strategies reported by the agencies include banning all motorized boats, not allowing Personal Watercraft (PWCs) or two-stroke engines, setting speed limits, restricting access to specific locations within the reservoir, and setting up or upgrading treatment facilities.

Information on current marine engine technologies was also collected to provide input data to the model and help guide reservoir managers' decisions on appropriate engine types to allow in reservoirs.

Donner Lake Test Site

Donner Lake is situated in the Sierra Nevada, close to Lake Tahoe in northern California, and receives the majority of its water supply from snowmelt. During the summer months, recreational boating, including fish boats, ski boats, and jet skis, occupies the water, with approximately 60 percent of the boats powered by two-stroke engines. Data from extensive, long-term limnological studies on the lake were incorporated into the study, and a meteorological station was installed on the lake during the year 2000 boating season for the purposes of this project.

A boat use census was conducted during the summer 2000 boating season to assess local use patterns and fuel use by engine type. Information was collected on the public boat ramp, concessionaires, and whole lake boating activities. A database was developed to store and organize the resulting data. Additionally, an MTBE and BTEX monitoring program was instituted during the summer 2000 boating season for the purpose of verifying the DLM-2002 predictions.

MODEL CALIBRATION AND VERIFICATION

As with any model, the reliability of DLM's output is only as good as the input data. In this case, special care was taken to obtain accurate boat usage data to ensure more accurate quantification of MTBE loading on a daily basis. Boat owners returning to the ramp were questioned as to what type of activity they had been engaged in, for how long and how much fuel they had used. This information was used to estimate MTBE loading for each month and time of week.

SPREADSHEET MODEL

UCD developed an elegant and easy to use spreadsheet program to estimate MTBE and BTEX concentration levels in reservoirs. This program is based on an extensive series of DLM-2002 model runs. The user inputs local reservoir data, chooses a compound, and the spreadsheet calculates a predicted maximum concentration for the boating season.

The estimator was tested on several reservoirs for which corroborating data exist. It was found to predict conditions with reasonable success, with the exception of very shallow reservoirs. If the reservoir is shallow enough that it is mixed to the bottom, the estimated volatile organic compound level may be underpredicted.

RESERVOIR MANAGEMENT GUIDANCE

MTBE management options can be divided into three groups: source control, treatment, and operational management. Source control options include banning recreational boating or restricting boating based on numbers or types of engine. Treatment options involve using advanced technologies to remove MTBE or BTEX from the water. Operational management includes changing the location of the drinking water intake in the reservoir to avoid intake of levels higher in MTBE content.

DLM-2002 was used to quantify these proposed management options. The results clearly showed that the proposed boating management options are highly effective in lowering MTBE levels.

Prior to deciding what option is most appropriate, it is necessary to weigh the costs of implementing an option against the loss or potential loss of economic activity that may result. Recreational boating contributes significant economic activity in those regions where it is enjoyed.

Both direct costs to the agency or water utility and regional economic (indirect) costs need to be considered. Direct costs of reducing boating activity will primarily affect water utilities; these include monitoring, enforcement, and compliance costs. Items contributing to this cost would typically be boats, enforcement personnel, gates, and security. In cases where water treatment or intake management are adopted, the direct costs are the amortized capital and annual operating costs of the facilities that will need to be constructed and operated to implement those options. Regional economic impacts would occur due to loss of recreation-related economic activity.

Banning all boating activity was obviously most effective but this option carries with it a high impact on the local economy. Less restrictive options such as engine type restrictions or various boating density restrictions were very effective in reducing MTBE levels with a much lower adverse economic impact.

Once MTBE has entered a water body, the most promising treatment technologies include air stripping, advanced oxidation, carbon adsorption, and synthetic resin sorbents. Treatment costs depend on existing system conditions, such as flow rates, MTBE influent concentrations, and the level of treatment needed to reach the desired effluent goal. The costs of the treatment options are much higher than controlling contaminant levels through a source control option. Controlling MTBE by varying the location of the new water intake, while less expensive than the treatment option, is more expensive than any of the management options.

CHAPTER 1
INTRODUCTION, BACKGROUND, AND PROJECT OBJECTIVES

INTRODUCTION AND BACKGROUND

This chapter describes the history, characteristics, and regulatory setting of methyl tertiary butyl ether (MTBE) and benzene, toluene, ethylbenzene, and xylenes (BTEX). These compounds can contaminate water supplies via motor vehicle fuels. In particular, the main focus here is on the transmission of these chemicals to drinking water reservoirs by recreational boating. This transmission is a cause for concern because of the potential health and aesthetic effects of these compounds and their current regulatory status. BTEX constituents are natural components of fuel with proven adverse potential effects on human health. BTEX levels in drinking water are regulated by the United States Environmental Protection Agency (USEPA). MTBE, a fuel additive used to reduce air emissions, has been classified as a potential human carcinogen. Although there is currently no primary drinking water standard for MTBE, the USEPA is expected to propose a secondary standard. Additionally, some states have established drinking water standards for MTBE.

Contamination of ground and surface waters by motor vehicle fuels and fuel additives is not a new problem, given the pervasive use of fuels in the past century. However, the discovery of MTBE in groundwater, lakes, and reservoirs used for drinking water has focused attention on the somewhat different characteristics associated with fuels containing additives known as oxygenates. Oxygenates are chemicals added to gasoline to increase the amount of oxygen in formulated fuel, thus reducing unburned hydrocarbons. The most common oxygenates are ethers and alcohols.

One of the most commonly used oxygenates is MTBE. Although MTBE and, to a lesser extent, ethanol, currently dominate the marketplace, several other ethers and alcohols also may serve as oxygenates and could become more prevalent, depending on cost, ease of production, blending characteristics, and regulatory requirements. These oxygenates include ethyl tertiary butyl ether (ETBE), tertiary amyl methyl ether (TAME), tertiary amyl ethyl ether (TAEE), disopropyl ether (DIPE), dimethyl ether (DME), and tertiary butanol (TBA).

Widespread use of MTBE in gasoline as a fuel oxygenate and BTEX as a natural component of gasoline has resulted in contamination of drinking water sources. A growing number of studies have detected MTBE in water sources (groundwater and surface water) throughout the country; in some instances these contaminated waters are sources of drinking water (McCord et al. 1998). Low levels of MTBE affect the organoleptic properties of drinking water, while higher levels can have aesthetic and long-term health effects. This has raised numerous concerns about the safety of drinking water. Some of the major sources of MTBE and BTEX include leaking underground storage tanks (USTs) and pipelines, spills, and emissions from marine engines into lakes and reservoirs, and, to some extent, air deposition. Regardless of the ultimate regulatory fate of MTBE as a fuel oxygenate, it is clear that oxygenates, in general, will remain in use for their significant air quality benefits. As such, they will pose challenges to water utilities and agencies that are responsible for managing water quality in lakes and reservoirs.

The focus of the current study is MTBE and BTEX contamination and the management of these compounds in lakes and reservoirs, particularly those that serve as drinking water sources. Prior to this study, options to reduce MTBE and BTEX contamination in lakes and reservoirs have not sufficiently evaluated to give practical data and advice to water managers who wish to consider the benefits of using a lake/reservoir as a recreational and drinking water quality source, without introducing unacceptable health risks into the lakes and reservoirs being managed. The present study has been undertaken to develop such guidance.

This study is based on a rigorous model that has been fully calibrated by field monitoring. It is applicable to all water utilities and water management agencies with large surface water storage facilities—lakes and reservoirs—that either currently allow, or are considering the use of, recreational boating, but who wish to be able to do so without creating unacceptable health or aesthetic risks from MTBE, BTEX, or other fuel-related contaminants.

History of MTBE and BTEX in the U.S.

Oxygenates have been added to fuel to create oxygenated fuel since the late 1970s to increase the octane of motor fuel when lead was phased out. Oxygenated fuels are of several types – Oxyfuel, reformulated gasoline (RFG), and California Air Resources Board (CARB) phase 2 fuel — each with a different specification for oxygen content. Oxyfuel is mandated in carbon monoxide (CO) non-attainment areas and requires 2.7 percent oxygen by weight, which corresponds to 15 percent by MTBE volume. RFG is mandated in ozone non-attainment areas and requires 2 percent oxygen by weight, corresponding to 11 percent MTBE by volume. Finally, CARB phase 2 fuel is a California-only fuel based on a predictive model that requires 1.8 to 2.2 percent oxygen by weight.

The Clean Air Act Amendments of 1990 created two fuel programs that required the addition of oxygenates to fuel. The first program, the Winter Oxyfuel Program, was implemented in 1992. It requires that Oxyfuel be sold during the cold weather season for areas that failed to attain air standards for carbon monoxide (CO). Ethanol is the primary oxygenate used in this program. The second program, the Year–round Reformulated Gasoline Program, requires the use of RFG in selected areas having the highest levels of tropospheric ozone. According to USEPA estimates, about 30 percent of the gasoline in the United States (U.S.) is RFG, of which about 87 percent contains MTBE (USEPA 2001).

To achieve specified oxygen content, approximately 15 percent MTBE by volume or 7.5 percent ethanol by volume can be used to yield the 2.7 percent oxygen by weight for the winter fuel program, and approximately 11 percent MTBE by volume or 5.5 percent ethanol by volume for the 2 percent oxygen required for RFG. Gasoline containing 10 percent ethanol, often referred to as gasohol, represents about 10 percent of all gasoline sold in the U.S.

Current Concerns

MTBE currently dominates the marketplace to meet the oxygen requirements mandated in the Clean Air Act Amendments of 1990. This is because of its low cost, ease of production, and favorable transfer and blending characteristics. In 1996, MTBE was the fourth largest produced organic chemical after ethylene, propylene, and ethylene dichloride, providing 17,075 million pounds (McCord et al. 1998). Currently, MTBE is found in 84 percent of all RFG, which constitutes over 30 percent of all gasoline. Since RFG is used in most urban areas, it is estimated that 40 percent of the U.S. population live in counties where MTBE is an additive (USEPA 1997).

BTEX also comprises a significant percentage of petroleum products. Typically, gasoline is composed of various hydrocarbons; the main compounds are alkanes, monocycloalkanes, dicycloalkanes, akylbenzenes, indanes, tetralins, naphthalenes, and some oxygenated alcohol additives. BTEX constitutes the alkylbenzenes and contributes about 18 percent weight by weight (w/w) in a standard gasoline blend. It must be noted, however, that the composition of gasoline can vary greatly depending on the manufacturer, the refining process, and the time of production. Of the various components in gasoline, BTEX is the largest group that has been related to adverse health effects.

Although air quality and economic benefits can be derived from the use of RFG with MTBE, such as reduced exhaust emissions of certain toxic compounds, discovery of MTBE in groundwater, lakes, and reservoirs used for drinking water has raised considerable concern among health officials and water suppliers. Because of MTBE's aesthetic and potential health effects and the fact that it is highly soluble in water and difficult to biodegrade, its persistence in water supplies has recently raised numerous research and management questions. In a nationwide survey of the occurrence and possible sources of MTBE and other volatile organic compounds (VOCs) in groundwater, the U.S. Geological Survey's (USGS) National Water-Quality Assessment program reported MTBE as the second most frequently detected chemical in shallow groundwater in urban areas (Reuter et al. 1998).

The use of MTBE in RFG has become a pervasive problem in California. It is reported that California is the third largest gasoline consumer in the world, with more than 13.7 billion gallons of gasoline used per year. As discussed in the previous section, because of air pollution problems, CARB and USEPA mandated the use of oxygenated fuels in California to improve vehicle octane ratings and reduce tailpipe emissions. Although MTBE is credited with significantly reducing air emissions from motor vehicles, it has also contaminated water supplies through accidental gasoline releases at dispensing sites, leaking product pipelines, and leaks from USTs (California MTBE Research Partnership 2000).

Monitoring data from November 1999 in California show that MTBE concentrations greater than the detection level have impacted 62 of the 6,409 drinking water sources sampled. However, contamination is not limited to California. Across the U.S., there have been numerous reports of MTBE leaks and spills from petroleum facilities, including refineries, terminals, pipelines, and service stations.

Consequently, MTBE has received widespread attention from municipalities and other drinking water providers, particularly in California, where major drinking water supplies have been significantly impacted by MTBE releases into the environment (California MTBE Research Partnership 2000).

BTEX compounds are less water-soluble than MTBE, but are acutely toxic and have noticeable health effects at high concentrations. Of greater concern in the current discussion, persistent exposure even to low concentrations of BTEX compounds can have long-term health effects.

Benzene is perhaps the most threatening of the BTEX constituents. Benzene has long been recognized as a human carcinogen, with chronic exposure linked to leukemia. Benzene exposure can also have other deleterious effects on the bone marrow and blood. It can decrease the number of red blood cells, leading to anemia, or cause excessive bleeding and increased risk of infection.

MTBE/BTEX Sources, Transport, and Exposure Pathways

MTBE enters the air through engine exhaust, evaporative emissions from incomplete combustion, and fumes released at refueling stations. Much of the interest in possible MTBE washout from the atmosphere to groundwater stems from recent publications by the USGS (Pankow et al. 1997; Baehr et al. 1999). These studies report that the possible sources for levels of MTBE found in shallow groundwater are primarily leaking USTs and pipelines; stormwater runoff and precipitation can also contribute small amounts of MTBE to shallow groundwater.

The sources that increase concentrations of MTBE also contribute to BTEX contamination of water and air. One of the most common sources of BTEX contamination in soil and groundwater is spills involving the release of petroleum products, such as gasoline and diesel fuel from leaking oil tanks. Because of their polarity and very soluble characteristics, the organic chemicals (BTEX) of petroleum products will be able to enter the soil and groundwater systems and cause serious pollution problems.

The discharge of unburned fuels from recreational motor boats and USTs has been identified as the primary sources of MTBE and BTEX in surface water (Kalman and Lund 1998). Use of gas-powered, aquatic recreational vehicles, specifically two-stroke personal watercraft

(PWC), which run on an oil/gasoline mixture, appear to be by far the largest source of MTBE in surface waters. These devices can release 4 to 50 percent of their fuel directly into the water, unburned (Juttner et al. 1995; Reuter et al. 1998; Malcolm-Pirnie 1998).

Relative to BTEX, two primary factors account for the buildup of MTBE in lakes and reservoirs during the boating season. First, MTBE is present in gasoline as a higher percentage than the individual BTEX compounds; and second, MTBE has a much lower Henry's law constant[2] than BTEX (about 14 times lower than that of benzene), which signifies that MTBE has a greater tendency than BTEX compounds to wash out of the atmosphere in precipitation and settle in the soil surface.

Sorption and dissolution are important mechanisms affecting fate and transport of BTEX in soil and groundwater systems. The sorption is controlled by characteristics such as the solubility, polarity, and octanol-water partition coefficient (Kow) of the BTEX compounds as well as the characteristics of the soil matrix and the fluid media. The BTEX compounds tend to be sorbed on the soil surface rather than to the soil matrix.

Due to its higher water solubility, MTBE exhibits lower adsorption to soil, and is more resistant to chemical degradation than BTEX. Its water solubility of about 50 gallons per liter (g/L) makes it about 25 times more soluble than benzene, the most soluble of the BTEX constituents. This implies that MTBE should be more readily leached to groundwater, and transported more quickly and further in groundwater, than the BTEX compounds. This is further supported by the fact that MTBE is more resistant to degradation than benzene, which would tend to allow its transport for longer distances prior to being attenuated via degradation.

The process of volatilization greatly determines the concentration of BTEX in the environment due to its high solubility, relatively low molecular weight, and high vapor pressure. (Benzene is the compound that volatilizes the easiest.) Oxygenates, such as MTBE, however, generally are less volatile than other major organic components of gasoline, such as BTEX compounds. Oxygenates, therefore, are more persistent than BTEX in the surface water environment.

MTBE's particular physico-chemical properties, such as high solubility, low soil-to-water partition coefficient, and low Henry's constant, make it somewhat difficult to remove from water

---

2 The Henry's law constant determines the tendency of a chemical to transfer from air to water, or vice versa.

(California MTBE Research Partnership 2000). Table 1.1 compares the physical and chemical properties of BTEX compounds and MTBE. As shown in Table 1.1, MTBE is significantly more soluble in water than are the BTEX compounds. This has implications in its fate and transport in lakes and reservoirs. Figure 1.1 presents the sources of MTBE and its fate and transport in air, surface, and groundwater. The surface water route is presented in bold, since the focus of this study is MTBE movement in lakes and reservoirs. Similar routes can be expected for BTEX constituents.

Table 1.1

Physical and chemical properties of BTEX compounds and MTBE

Physical and Chemical Properties	Benzene	Toluene	Ethylbenzene	*o*-xylene	MTBE
Molecular weight	78.11	92.14	106.17	106.17	88.15
Vapor Density @ 1 atm; 10°C	3.36	3.97	4.57	4.57	3.80
Henry's Law Constant (mol/kg*bar at 25°C)	0.16	0.15	0.12	0.19	1.6
Specific gravity @ 25°C	0.88	0.8669	0.867	0.8802	0.744
Boiling Point (°C)	80.1	110.6	136.25	144.4	53.6-55.2
Water Solubility (mg/L)	1,730	534.8	161	175	43,000-54,300 50,000
Vapor pressure (mm Hg) @ 25°C (kPa) @ 100F	7,695.19 1.26	28.4 37.9	9.53 1.23	6.6 0.889	245-276 3.33
Log Koc	1.18-1.99 1.50-2.16	1.56-2.25	2.94 1.98-3.04	1.68-1.83	1.091, 1.035, 1.049
Log Kow	2.36	2.73	3.24	3.10	1.20

(continued)

Table 1.1 (continued)

Physical and Chemical Properties	Benzene	Toluene	Ethylbenzene	*o*-xylene	MTBE
Diffusivity meter squared per second (m2/s)					
Liquid	9.95x10-10	8.87x10-10	8.6x 10-10	8.6x 10-10	8.58x 10-10
Gas	9.36x 10-6	8.35x 10-6	7.59x 10-6	7.56x 10-6	8.45x 10-6
Biodegradability	Aerobic	Anaerobic / Aerobic	Aerobic	Aerobic	Aerobic

Oxygenates can be transformed through a variety of chemical, physical, and biological processes. In surface water, photolysis is the most important transformation process for ether oxygenates (e.g., ETBE, TAME, TAEE, DIPE, DME), while biodegradation is more important for alcohols. Biodegradation does occur for ether oxygenates, however. Tertiary butanol has been shown to be a major byproduct of MTBE biodegradation. Ethanol may pose a different issue with respect to biodegradation. Recent reports have noted that the presence of ethanol in gasoline may inhibit the biodegradation of BTEX compounds, perhaps because microbes preferentially metabolize the ethanol. This phenomenon may increase the adverse impact of BTEX compounds, making BTEX contamination more persistent.

MTBE has been found to be persistent relative to BTEX compounds; however, its biodegradation is actively being investigated by several researchers (e.g., Hanson et al. 1999; Salanitro et al. 1999). MTBE degrades rather rapidly in the atmosphere. The primary degradation product is tertiary-butyl formate. Half-lives of MTBE in the atmosphere can be as short as 3 days. The primary reactant in the atmosphere responsible for the degradation is the hydroxyl radical. In urban areas, degradation may be lower due to lower concentrations of the hydroxyl radical being present. Nonetheless, MTBE is not thought to persist in the atmosphere, and measured concentrations result from a steady state between vehicular emissions and degradation in the atmosphere.

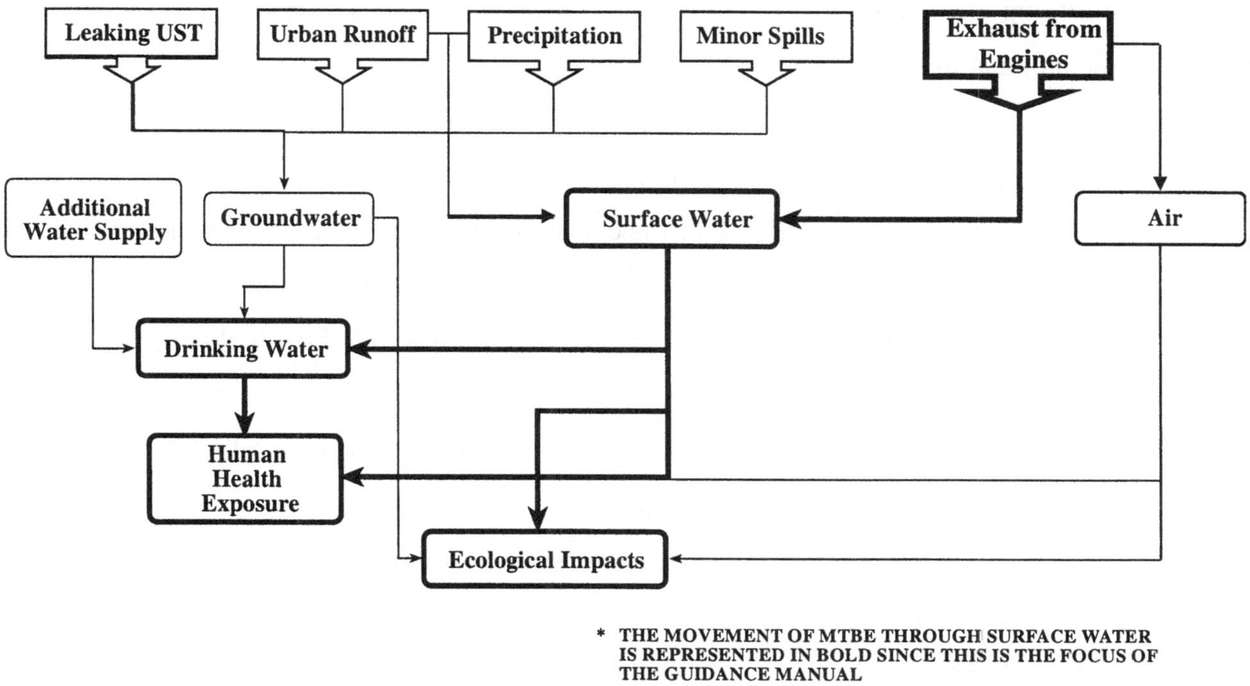

Figure 1.1 MTBE movement in the environment

MTBE degradation in soil and groundwater is under extensive investigation. Evidence to date indicates that MTBE degrades much slower than the BTEX compounds, and, therefore, may travel further and persist longer in groundwater plumes. The half-life of MTBE in water is estimated to be approximately 4 weeks (Rathbun 1998). The primary reaction product is tertiary butanol, which is another constituent of gasoline commonly found in conjunction with MTBE. The ultimate degradation products have been confirmed to be carbon dioxide and water (New Jersey Department of Environmental Protection 2000).

While natural field degradation of MTBE by microorganisms is rather slow, it is measurable, and particular species have recently been identified that are able to utilize MTBE as a sole-carbon source. Recent research in bioremediation of MTBE implies that degradation can be greatly accelerated by inoculating water or soil with MTBE-acclimated microorganisms. Thus, while natural degradation of MTBE may be slow, there is a good potential for using enhanced biodegradation as a remediation technique. Spontaneous development of higher degradation rates has also been observed in one California wastewater treatment plant. Factors controlling the rate of biodegradation of MTBE appear to be site-specific, and are not well understood at this time.

The degradation of BTEX compounds is another aspect that affects the concentration in soil and groundwater. The natural bacterial flora in soil has an ability to aerobically degrade the BTEX. This degradation will reduce the concentration in soil and groundwater. Anaerobic biodegradation has been repeatedly established when dealing with toluene, and in some cases, with xylenes. As shown in Figure 1.1, exposure pathways to MTBE include air, surface, and groundwater. Ecosystem exposure is predominantly through lake recreational contamination, while human exposure is typically through lake and groundwater-supplied drinking water, or combined surface and groundwater supplies.

BTEX compounds are volatile and soluble in water. Ingestion of or dermal contact with contaminated water and inhalation of vapors can be important exposure pathways for BTEX compounds. Because of its polarity and soluble characteristics, BTEX readily enters various media, such as groundwater and soil systems, and are considered a serious problem due to their acute and long-term toxic effects. In addition to toxicity, benzene is a known carcinogen.

Regulatory Setting

The Safe Drinking Water Act (SDWA) is the main federal law that ensures the quality of drinking water in the U.S. It was originally passed by Congress in 1974, and amended in 1986 and 1996. Under SDWA, USEPA is authorized to set standards for drinking water quality. USEPA also oversees the states, localities, and water suppliers that implement those standards.

There are two categories of drinking water standards: a National Primary Drinking Water Regulation (NPDWR or primary standard) is a legally- enforceable standard that applies to public water systems. Primary standards protect drinking water quality by limiting the levels of specific contaminants that can adversely affect public health and are known or anticipated to occur in water. Primary standards initially take the form of Maximum Contaminant Levels (MCL), which is the maximum permissible level of a contaminant in water that is delivered to any user of a public water system. A National Secondary Drinking Water Regulation (NSDWR or secondary standard) is a non-enforceable guideline regarding contaminants that may cause cosmetic effects (such as skin or tooth discoloration) or aesthetic effects (such as taste, odor, or color) in drinking water. USEPA recommends secondary standards to water systems but does not require systems to comply. However, states may choose to adopt them as enforceable standards.

USEPA does not regulate all contaminants that are emitted by recreational watercraft. However, BTEX and MTBE are regulated at the federal and state level. USEPA has developed maximum contaminant levels (MCLs) for BTEX. USEPA has not set a national primary drinking water standard for MTBE. Appendix A (Table A.1) lists the VOCs emitted by recreational watercraft and, if applicable, the corresponding MCL and MCL Goals (MCLGs) set by USEPA.

Although USEPA has no primary drinking water standard for MTBE, MTBE has an objectionable taste and odor, so USEPA has set a drinking water advisory of 20 to 40 μg/L for the contaminant (USEPA 1997). In addition, as part of implementing the SDWA Amendments of 1996, USEPA placed MTBE on the drinking water Contaminant Candidate List (CCL) for further evaluation to determine whether or not a NPDWR or NSDWR is necessary. The CCL divides contaminants into the following categories:

- Those that are priorities for additional research
- Those that need additional occurrence data
- Those that are priorities for consideration for rulemaking

In September 1999, USEPA revised the Unregulated Contaminant Monitoring Rule (UCMR) under the SDWA requiring all large public water systems as well as a representative sample of small systems to monitor for MTBE in groundwater and surface water, beginning in 2001. The rule includes requirements to submit the monitoring data to USEPA for inclusion in the National Drinking Water Contaminant Occurrence Database, and requirements to notify consumers of the results of monitoring. USEPA determined that MTBE needs more health effects research and occurrence data before a regulatory determination on MTBE can be made (USEPA 2000; USEPA 2001b).

As an additional interim measure, USEPA issued a health advisory in December 1997 entitled, "Drinking Water Advisory: Consumer Acceptability and Health Effects Analysis on Methyl Tertiary Butyl Ether (MTBE)." The health advisory provides guidance to communities that could become exposed to drinking water contaminated with MTBE. The advisory recommends control levels (i.e., 20 to 40 parts per billion [ppb]) to prevent adverse drinking water taste and odor (USEPA 2000).

USEPA was expected to propose a secondary drinking water standard in spring 2001, which was to be based on taste and odor (USEPA 2001). The standard has not yet been proposed.

State Regulations

California has listed MCLs for all four BTEX constituents. The California Department of Health Services (CA DHS) set the Benzene MCL at .001 milligrams per liter (mg/L). The California MCL for ethlybenzene is identical to the federal MCL at 0.68 ppm. Several states have also set MCLs for ethlybenzene. Arizona, Kansas, Maine, Minnesota, Rhode Island, and Wisconsin have set their levels at about the same concentration whereas Illinois has set its MCL at 0.001 ppm. The California MCL for toluene is also identical to the federal MCL at 0.15 mg/L. For xylene, California adopted an MCL of 1.75 ppm in 1987. CA DHS also proposed a secondary maximum contaminant level (SMCL) for MTBE at 5 μg/L based on available data for observable detection thresholds. As is often true for California environmental protection standards, this proposed SMCL is considerably more stringent than USEPA's drinking water advisory.

Several states across the country have adopted, or are in the process of adopting, standards or guidance values for drinking water. Six states have adopted their own primary standard for MTBE in drinking water: California, Maine, Michigan, New Hampshire, and New Jersey. While some states have adopted MCLs, others have guidance values in place on the basis of USEPA's Health Advisory Program.[3] Delaware is considering an MCL for MTBE of 0.01 ppm (Beckel 2001). Table 1.2 presents the status of groundwater and drinking water standards for MTBE in several states across the country.

[3] USEPA's Health Advisory Program was initiated to provide information and guidance to individuals or agencies concerned with potential risk from drinking water contaminants for which no national regulations currently exist. Advisories are not mandatory standards for action. Advisories are used only for guidance and are not legally enforceable. They are subject to revision as new information becomes available. USEPA's Health Advisory Program is recognized in the Safe Drinking Water Act Amendments of 1996. The health advisory includes consumer acceptability advice, as "appropriate", under this statutory provision, as well as a health effects analysis.

Table 1.2

State agencies with MTBE drinking water standards*

State	Responsible State Departments	Groundwater / Drinking Water Standard	Reference(s)
Arizona	Arizona Department of Environmental Quality	Health-based guidance level 35 ppb (unenforced)	DEQ 2001
California	California Department of Health Services	Primary MCL†: 13 ppb Secondary MCL‡: 5 ppb (enforced)	CA DHS 2001
Connecticut	Connecticut Department of Public Health	70 ppb	Hitzig et al. 1998
Idaho	Idaho Department of Health Welfare	Guidance level 261 ppb - 511 ppb depends on use of groundwater	DHW 2001
Maine	Maine Department of Environmental Protection (ME DEP) Maine Department of Human Services, Bureau of Health	35 ppb	Mower 2001 ME DEP 2001
Massachusetts	Massachusetts Department of Environmental Protection	Guidance Level 70 ppb	MA DEP 2001
Maryland	Maryland Department of the Environment	20 ppb	MA DE 2001
Michigan	Michigan Department of Environmental Quality	240 ppb	MI DEQ 2001

Notes:
* The information in this table is correct as of July 2001
† Maximum Contaminant Level (MCL)
‡ The secondary MCL is based on taste and odor concerns

(continued)

Table 1.2 (continued)

State	Responsible State Departments	Groundwater / Drinking Water Standard	Reference(s)
North Carolina	North Carolina Department of Environment and Natural Resources	200 ppb	NC DENR, 2001
New Hampshire	New Hampshire Department of Environmental Services (NH DES) New Hampshire Department of Health and Human Services (NH DHHS), Bureau of Health Risk Assessment (BHRA)	13 ppb	Talbot 2001 NHDES 2001 NH DHHS 2001
New Jersey	New Jersey Department of Environmental Protection: Division of Science, Research, and Technology	70 ppb	NJ DEP 2001
New York	New York State Department of Environmental Conservation: Division of Water Resources	50 ppb	NY DEC 2001
Pennsylvania	Pennsylvania Department of Environmental Protection	20 ppb	PA DEP 2001
Rhode Island	Rhode Island Department of Health	Health based guidance level 40 ppb	RI DH 2001
Vermont	Vermont Department of Environmental Conservation	Health Advisory Standard (enforced) 40 ppb	VT DEC 2001
Wisconsin	Wisconsin Department of Natural Resources	60 ppb	WI DNR 2001
Wyoming	Wyoming Department of Environmental Quality	Guidance Level 40 ppb	WY DEQ 2001

Notes:
* The information in this table is correct as of July 2001
† Maximum Contaminant Level (MCL)
‡ The secondary MCL is based on taste and odor concerns

REVIEW OF PREVIOUS STUDIES ON MTBE

Related studies on MTBE range from those focusing on predictive models of MTBE's fate and transport to those reviewing and evaluating management options for human and ecosystem impacts. This review focuses on relevant, notable, previous studies carried out in California.

A comprehensive evaluation of management options to minimize water supply and ecosystem impacts of MTBE was carried out by Kalman and Lund (1998) at UCD. The study was a policy review and evaluation for MTBE management for the California Legislature; no field work, surveys, or analyses were conducted. The 1998 study evaluated policies for MTBE management using five scenarios developed to account for water uses, type of water source, and users. The five scenarios were: (1) lake recreation and ecosystem exposure, (2) lake-supplied public drinking water, (3) groundwater-supplied public drinking water, (4) diversified public drinking water (combined groundwater and surface water supplies), and (5) groundwater-supplied, rural-household drinking water.

The 1998 study compared MTBE-control policies for both human health and ecological risk objectives. The MTBE-control policies included to reduce human health exposure were based on: (1) reduction in MTBE concentration, (2) reduction of expected exposure, and (3) economic cost effectiveness. MTBE concentration of surface and groundwater (in μg/L or ppb) with and without proposed policies were compared to USEPA proposed secondary standard of 5 ppb and rated based on likelihood of exceeding this proposed standard. The results of the 1998 study were used to deduce that groundwater contamination by MTBE is orders of magnitude higher than surface water contamination. Therefore, human health is more likely to be affected by groundwater contamination than by surface water contamination.

The findings of the 1998 study indicated that ecological health, contingent on surface water, did not appear to be affected by MTBE at the levels produced by boating activities at the two lakes studied (Lake Perris and Donner Lake). Of the alternative policies reviewed, the 1998 study concluded that treatment appeared to be the most cost-effective for groundwater contaminated by MTBE. For surface water, the study concluded that reduction in two-stroke-engine boating would reduce MTBE levels to acceptable concentrations; this option was more cost-effective than other alternatives, such as treatment. In the short term, the study concluded that the cost of currently available oxygenated fuels were comparable. In the long term, the costs of using either ethanol oxygenated fuel or non-oxygenated fuel could be lower than using MTBE-oxygenated fuel. The focus of the 1998 study was entirely on MTBE; it did not discuss BTEX in the environment or its management in various media.

Reuter et al. (1998) studied the fate and transport of MTBE in Donner Lake, which is a multiple–use lake. The purpose of this study was to determine the (1) relative contribution of motorized watercraft as a source of MTBE, (2) its seasonal distribution, (3) loss from the water column, (4) extent of vertical transport, and (5) persistence between years. MTBE measurements were made at 9 individual depths from surface to bottom on 16 dates. Recreational boating was observed to be the most important source of MTBE. Statistically, 86 percent of the change in MTBE was explained by variation in motorized watercraft use. Neither highway runoff nor precipitation was considered to have contributed significantly to MTBE levels in the lakes and reservoirs. The major loss of MTBE appeared to be volatilization at the air-water interface. Thermal stratification acted to retard MTBE transport to deeper depths.

A few studies have used predictive models to understand the behavior of MTBE in surface waters. Most of these studies have focused exclusively on representing volatilization (Pankow et al. 1996). Moreover, these studies have focussed on rivers and streams (Fischer et al. 1979), which are generally well mixed vertically, unlike lakes and reservoirs where there is density stratification and lower turbulence intensities. This significantly affects the fate and transport of MTBE and other contaminants such as BTEX.

Studies, such as those by Anderson (1997) and Malcolm-Pirnie (1998), that have used lake models have neglected vertical changes in the surface layer depth and selective withdrawal processes. Such studies can only work in the simplest of cases. These studies did not model the key mixing processes, or describe the seasonal variations in the vertical stratification of a reservoir throughout the year and through sequences of multiple years. A dynamic lake model coupled with an MTBE fate model may be used to explore a range of management options, which will consider both volatilization of MTBE across the air-water interface and MTBE influx from recreational boat use only.

Recognizing these limitations, McCord and Schladow (1998) used a dynamic lake model coupled with an MTBE fate model to study the transport and fate of MTBE in lakes and reservoirs. The study indicated that the combined model might be used to explore a range of management options. A more detailed discussion of this modeling effort is presented in Chapter 2. Results of this effort in Donner Lake gave predicted MTBE concentrations that followed observed patterns but not the exact values. The study concluded that greater functionality in emission rates, requiring more accurate data (boat type and operation time), would likely improve model results.

PROJECT OBJECTIVES

Most of the previous studies have focussed primarily on MTBE, with limited discussion of the fate, transport, or management of BTEX. Because BTEX and other polyaromatic hydrocarbons (PAHs) also pose potential health risks and can serve as indicators of specific contaminants of future interest, gathering data and transport characteristics of these compounds in relation to MTBE was determined by the authors and sponsors of this study to be very useful and relevant to the industry regardless of the regulatory future of MTBE. These gaps in the previous policy evaluation and fate- and transport-related studies, as well as modeling efforts, led to the design of the objective and the approach of the present research. This research was therefore conceived and designed to ensure that all the information available on BTEX and MTBE – including their chemistry, source, fate, and transport and management options – together with the predictive combined model, be used to develop effective strategies for their management.

A number of methods are available for reducing fuel emissions in lakes and reservoirs. These include certain operational strategies, such as altering the frequency and/or depth of water withdrawal and selective management of boating frequency and density. Furthermore, new boat-motor technologies, such as two-stroke engines modified with direct fuel injection, can reduce fuel and MTBE emissions. As discussed above, none of these options has been sufficiently evaluated to give practical data and advice to water managers who wish to use these methods to obtain the benefits of dual use without introducing unacceptable aesthetic or health risks. The primary objective of this project, therefore, is to develop a guidance manual for water utilities and agencies to fill this data gap.

As discussed above, USEPA considers MTBE to be a potential human carcinogen. Some water management agencies have either severely restricted or eliminated the use of recreational watercraft from source waters and reservoirs to preclude the possibility that MTBE will threaten the health of their water supplies. Many other agencies are actively debating their most appropriate course of action. Prior to this study, research-based guidance to water suppliers was not readily available to inform the debate.

The study also rigorously evaluates options available to water suppliers to reduce MTBE and other fuel-related contaminants to acceptable levels in their source waters, while still allowing recreational boating opportunities. Techniques evaluated include the location and timing of water withdrawals; the use of new boat-motor technologies that reduce the amount of fuel discharges; and the management of the type, frequency, and density of boating allowed. These techniques are compared to the cost and difficulty of treating the raw water for removal of MTBE.

The ultimate product of this research is a sophisticated, one-dimensional MTBE/BTEX fate and transport model and guidance directed to water agency managers. The model gives practical, balanced advice on alternative means to adequately protect water supplies while supporting recreational opportunities. A spreadsheet model, which is based on an extensive set of runs with DLM-2002, accompanies this report, for use by managers of water supply lakes and reservoirs.

Project Significance and Applicability

The accumulated knowledge on MTBE distribution, volatilization, emission rate as a function of engine type, etc., and the ability to model MTBE change based on boat use and water withdrawal can be used to help water suppliers develop strategies to maintain fuel-related contaminants at acceptable levels. The potentially conflicting mandate to supply safe and aesthetically acceptable drinking water, and to provide an opportunity for recreational activities on lakes and reservoirs used for drinking water, needs immediate resolution.

At this time, the ultimate fate of MTBE as a gasoline additive is unknown. Even if, as recommended in the University of California's Final Report from SB 521[4], MTBE were to be phased out over the course of several years, agencies charged with supplying safe and acceptable

---

[4] SB 521 refers to California Senate Bill No. 521 (1997), which requires study and evaluation of impacts from MTBE on public health and the environment. Pursuant to SB 521, the University of California, Davis, carried out scientific studies on MTBE.

drinking water need advice to define the most practical interim measures. Reservoir operators outside of California will face similar issues well into the future and will especially benefit from this study.

Other fuel-related contaminants, such as BTEX and PAHs, also pose potential health risks and can serve as indicators of specific contaminants of future interest. The gathering of data and transport characteristics of these compounds in relation to MTBE would be very useful and remain relevant to the industry regardless of the regulatory future of MTBE.

Within California alone, there are more than 100 large-surface lakes and reservoirs used to store drinking water supplies, and there are hundreds more throughout the U.S. Most of these lakes and reservoirs are publicly owned, some of them are already prime recreational attractions with strong economic and quality-of-life impacts on their communities, and many others offer the potential for recreational boating activity, for which there is always strong public pressure. The threat of contamination from fuel-related contaminants is causing many water management agencies and utilities to either prohibit or severely restrict recreational watercraft from their lakes and reservoirs.

New developments in marine engine technology can reduce fuel emissions into lakes and reservoirs, and operational and boating management options can significantly reduce levels of fuel-related contaminants. This study provides data and advice to water managers who wish to consider the benefits of dual use without introducing unacceptable aesthetic and/or health risks into lakes and reservoirs. The research is based on a rigorous model that has been fully calibrated by field monitoring.

This research will thus be applicable to all water utilities and water management agencies with large-surface water storage—lakes and reservoirs—that either currently allow, or are considering the use of, recreational boating, and wish to do so without creating unacceptable aesthetic and/or health risks from fuel-related contaminants.

RESEARCH APPROACH

A multi-pronged research approach was adopted to develop this study for water utilities and water management agencies. The approach was designed to include sequential and highly interrelated tasks, including:

1. Conducting a survey of existing data on management options and marine engine technology: This included documenting the management options used for reducing MTBE and other fuel-related contaminants from various water utilities and water management agencies through a Reservoir Management Questionnaire. Results from the Reservoir Management Questionnaire were used to develop management scenarios for the UCD DLM-MTBE model. The questionnaire responses were also helpful in considering economic implications of the various management scenarios. The survey phase of the research also included a review of state-of-art marine engine technology, including data on operating principles and emissions. The engine technology data were used to develop management scenarios; emission data were also used as inputs to the UCD DLM-MTBE model.
2. Predicting MTBE and BTEX concentrations in lakes and reservoirs: This involved using a mathematical model (the UCD DLM-2002) to examine MTBE and BTEX concentration and distribution in reservoirs, and predicting their concentration and distribution using various management scenarios. Results from a boating census and detailed field monitoring were used to refine and calibrate the model to improve its ability to predict fuel-related water quality resulting from various management scenarios.
3. Evaluation of management scenarios: The information derived from the data survey and modeling activities was used to evaluate various management scenarios to present management options for utilities and agencies responsible for surface water systems. This study serves presents options to water suppliers and recreation managers for designing lake and reservoir management and operational strategies that protect drinking water quality while maximizing recreational capabilities.

The methodology adopted included questionnaire-based surveys, literature review of the existing engine technologies, mathematical modeling to predict and evaluate management scenarios, and field tests to validate and test the UCD DLM-MTBE model.

CHAPTER 2
MODELING MTBE AND BTEX

INTRODUCTION

This chapter describes the techniques used to model MTBE and BTEX, including a summary of previous modeling efforts. The model used for this research, the UCD Dynamic Lake Model, was developed at the University of California at Davis. It examines how MTBE and BTEX are distributed in reservoirs and how their concentrations can be controlled through management techniques. The model is based on an earlier series of reservoir models (Imberger et al. 1978; Imberger and Patterson 1981; Hamilton and Schladow 1996; Schladow and Hamilton 1996). DLM was first used in 1998 to model MTBE, and that version is known in this Guidance Manual as DLM-1998. Although the 1998 version was found to be scientifically sound, the model's authors felt that it was limited by the quality of the boating and engine emissions data. A more accurate accounting of boat traffic, in the form of the Donner Lake 2000 boating census, and improved data on engine emissions were therefore used in the modeling for this manual. In addition, several additions and improvements, as described below, were made to the model itself. The resulting version is referred to in this Guidance Manual as DLM-2002.

Predictive models of the concentration of MTBE and other VOCs in surface waters have generally focused on representing volatilization. This approach is considered inappropriate for lakes and reservoirs, where density stratification and lower-turbulence intensities prevail. Lakes and reservoirs in temperate climates typically undergo an annual cycle. During the spring and summer, lakes and reservoirs become density-stratified, with warm, lighter water residing as a layer at the surface, and cool, denser water at depth. This stratification inhibits vertical mixing. Therefore, dissolved substances (such as MTBE) introduced at the surface, will tend to remain in close proximity to the surface. In fall and winter, the surface layer of the lake cools and becomes denser, leading to its deepening as it mixes with the underlying water. In most cases, this deepening of the surface layer continues to the bottom of the lake, a process known as complete overturn.

The distribution of streams or other inflows within a lake are determined by the density distribution of the lake's water column. The depth at which stream water intrudes into a lake depends on the stream temperature and on the ambient stratification in the lake at the time of inflow. The density stratification also impacts the removal of water from lakes and reservoirs. The vertical distribution of water quality is constrained by the stratification pattern. Water managers typically attempt to target strata of higher quality water, or to avoid layers of undesirable water, a process known as selective withdrawal. The extent to which this is possible is governed by the existing infrastructure in a reservoir. Hydrodynamics in general, and vertical mixing processes in particular, have a controlling influence on the distribution of MTBE and BTEX. Selection of a model that can accurately represent these processes is therefore critical in evaluating the most effective management options for dealing with MTBE and BTEX contamination.

DYNAMIC LAKE MODEL (DLM)

The fate and transport model employed for this research is the UC Davis Dynamic Lake Model (DLM). The following discussion refers to both DLM-1998 and DLM-2002 (McCord and Schladow 1998). DLM is a one-dimensional simulation model of the vertical distribution of temperature and salinity in small- to medium-size lakes and reservoirs. The assumption of one dimensionality means that variations in density, temperature, and VOC concentration in the lateral directions are assumed to be small when compared with variations in vertical directions. The core of DLM is a hydrodynamic model that parameterizes the individual transport and mixing processes in a lake and sequentially couples them to predict the lake's density structure over time (McCord and Schladow 1998).

Model Structure

The vertical profile of the lake is represented as a set of up to 200 Lagrangian layers, which are free to move vertically and to contract and expand in response to inflows, outflows, and surface-mass fluxes. The Lagrangian formulation avoids the need to calculate vertical velocities, greatly decreasing computational time and minimizing numerical diffusion, as compared with a fixed-grid Eulerian approach. Each layer is homogeneous, and property differences between layers represent the vertical distribution. Layer thickness is adjusted within the model according to the resolution required to represent the vertical density gradient. The model is initialized with measured data for each layer.

Model Processes

Mixing is represented by the amalgamation of layers. Thus, for example, when wind mixing has produced sufficient turbulent kinetic energy to overcome the potential energy difference (density stratification) of the two upper layers, the two layers are simply combined. Properties of the amalgamated layer are volumetrically averaged, and the total number of model layers is decreased accordingly. Amalgamated layers may be split according to a specified maximum-layer thickness criterion, thereby maintaining the desired spatial resolution. Conversely, when a layer size falls below a specified minimum criterion, as may occur when there is withdrawal of water from the lake, then the layer is amalgamated with the smaller of the two bounding layers. Inflows are inserted at their level of neutral buoyancy after allowance for entrainment has been made. To model withdrawal of water from a lake, a withdrawal layer thickness and location are calculated based on the height and type of outlet port, the local density stratification at that height, and the rate of withdrawal.

Volatilization of fuel-based VOCs from the surface of a lake is modeled as a two-film diffusion process, involving resistance in both the air and water phases. Diffusion rates in both the surface layer of water and the overlying air mass are important, although the overall rate tends to be limited by the liquid-side transfer velocity under the normal range of temperatures and wind speeds. Transport of VOCs between computational layers is either by advection, if there is sufficient turbulent kinetic energy, or by diffusion for stably stratified regions of the water column. In DLM-1998, VOC influx is assumed to be solely a function of the number of boats using the lake or reservoir. In addition to the number of boats, DLM-2002 also considers the mix of engine types powering the boats, the emission rates of their engines, fuel consumption rates, the volume fraction of the compound in fuel, and length of time on the lake.

Model Inputs

Primary model input consists of seven files that are prepared before a simulation. These include files containing daily average values for meteorological data, reservoir inflow and outflow, and VOC loading; and files for initial conditions, reservoir area and volume, and physical constants.

In addition to the input files, there are a number of variables that can be set by the user when the model is run. All are entered through the reservoir management options sub-menu. Included are multipliers for wind speed, inflow, and outflow, so that these parameters can be changed dynamically, without having to revise the input files.

For DLM-1998, a file containing estimated daily boat counts and a per-boat MTBE load are supplied as input to the model. For DLM-2002, average MTBE loads by month and time of week are computed in a spreadsheet before the model is run. At runtime, the model uses boating activity data to compute the MTBE load for each day modeled. The average loads are computed assuming that all fuel contains MTBE. Since it is known that some of the fuel used at Donner Lake was MTBE-free, DLM-2002includes a user-entered MTBE multiplier, whose primary purpose is to specify the fraction of total fuel that contains MTBE.

Model Outputs

Primary model output consists of daily vertical profiles of temperature and VOC concentration in the lake, and temperature and VOC concentration of the outflowing (or withdrawn) water. Since the one-dimensional version of DLM was used in this research, no horizontal distribution of parameter values is available.

DESCRIPTION OF DLM-1998

DLM-1998 was developed for, and used in, the 1998 study by McCord and Schladow (1998). The hydrodynamic model was linked to an MTBE sub-model that accounted for volatilization of MTBE across the air-water interface, and considered MTBE influx only from two-stroke PWCs.

Simulations

Three specific lakes were simulated: Donner Lake in the Sierra Nevada, Lake Perris near Los Angeles, and Calero Reservoir near San Jose. These lakes were chosen because they represent a broad range of geographical, climatological, and limnological conditions. They are also three lakes for which sufficient data of the correct types were available to validate the hydrodynamic model and to calibrate the volatilization model. The model was employed to predict changes in MTBE concentrations for each lake and its outflow waters as a function of withdrawal conditions and boat use limits.

MTBE Volatilization Sub-model

Volatilization of MTBE from the surface of a lake was modeled as a two-film diffusion process, involving resistance in both the air and water phases (Lewis and Whitman 1924). Water-side transfer velocity was calculated following the pre-publication work of Bender et al. (2002). The air-side velocity formulation was that of Schwartzenbach et al. (1993).

Inputs and Outputs

Daily meteorological conditions, stream inflow, stream outflow (or reservoir withdrawals), and initial water quality conditions were used as input to the model. The model, as configured for the 1998 study, outputs daily vertical profiles of temperature and MTBE concentration in the lake, and temperature and MTBE concentration of the outflowing (or withdrawn) water. A more detailed description of the model is provided in Appendix B.

Results

Reduced input of MTBE (lower emission rates or boat use limits) resulted in decreased concentrations and decreased total mass of MTBE. Withdrawing water from deeper in the water column led to minimization of MTBE concentrations downstream, but only while the surface mixed layer remained above the intake level.

Results for MTBE concentrations in the three lakes followed observed patterns, but not the exact values. Boating data for DLM-1998 consisted only of a daily estimate of the number of two-stroke PWCs, and a single value for MTBE release from each boat. It was felt that greater functionality in boating data, including engine type, operation time, fuel use, and more accurate emission rates, would likely improve model results.

Model Review

This model was included in the UC SB-521 report delivered to the California Governor's Office in November 1998. As part of the study, this model was scientifically reviewed by a host of experts, including governmental agencies, outside water quality consultants, and, especially, the U.S. Geological Survey, which was specifically requested by the State of California to conduct a review of this work. At no time during this highly scrutinized process was the UCD MTBE model considered scientifically inappropriate or otherwise flawed. The model's authors stated, however, that the SB 521 version of the model was limited by input data and hence modifications were proposed.

MODIFICATIONS TO DLM-1998

Introduction

The modifications to DLM-1998 focused on both model structure and input data. The new model, DLM-2002, is based on MTBE characteristics because of the large body of data on MTBE in surface waters collected by UCD and other investigators in California. DLM-2002 is capable of considering the full range of engineering and management options for controlling the concentration of MTBE and other VOCs in reservoir withdrawal water. Such options include operational strategies, such as the location and timing of water withdrawals; the use of new boat-motor technologies that reduce the amount of fuel discharges; the use of alternative fuels; and the management of the type, frequency, and density of boating allowed on the reservoir or lake.

Model Structure

Two key structural improvements were made for in DLM-2002. First, the model was expanded to include the capacity to model the four BTEX compounds. Second, a degradation sub-model was employed when it became apparent that volatilization alone could not account for the low measured levels of MTBE and toluene encountered in Donner Lake.

Additional Volatile Compounds

The expansion of the model to include BTEX was straightforward. There are four quantities associated with each fuel-based VOC: molecular weight, molar volume, Henry's Law constant, and first-order degradation coefficient. Except for the degradation coefficient, these were already in place for MTBE, and the relevant variables were simply expanded to include the others. Molecular (molar) weight is a known quantity. Molar volume is calculated by dividing the molar weight by the density. The Henry's Law constant is a function of the surface water temperature, and is calculated using the data of Robbins et al. (1993). Degradation coefficients were estimated using data compiled by Howard et al. (1991).

The model can thus be easily expanded to consider a greater range of volatile compounds besides MTBE and BTEX, and can provide a consistent framework to examine the fate of various gasoline components and byproducts in the future. These components may include ethanol and other potential gasoline oxygenates, as well as PAHs.

Degradation Sub-model

In the process of calibrating DLM-2002, it became apparent that volatilization alone could not account for the measured levels of MTBE and toluene in Donner Lake. Modeled levels of these two compounds were consistently higher than measured levels. MTBE, while problematic, was less affected than toluene because of its greater resistance to degradation, but it was clear that some other loss mechanism was operating, most likely microbial degradation.

An initial assumption was made that the process followed first-order kinetics. Rate coefficients were chosen, based on Rathbun (1998), which follows the work of Howard et al. (1991). There is limited research in this area, and the published aerobic degradation rates are given as a range. For MTBE, for example, the half-life ranges between 4 weeks and 6 months. The high value from the given range for each compound was chosen initially, and this turned out to provide a good fit with the measured data for MTBE and toluene. Additional details of the degradation sub-model are provided in Appendix B.

Model Input Data

DLM-1998 was believed to be limited by the quality of the input data, particularly with respect to the amount of boat use and MTBE emission rates per boat. To overcome these two areas, the following modifications were proposed: (1) a more detailed accounting of boat traffic; and (2) more improved data on engine emissions through the use of tank tests (C. Emmett, CARB) and in-lake experimentation.

Use of Boating Census Data

Chapter 3 describes the design and implementation of the boating use census at Donner Lake, and development of a database from the survey data. Information on the type and length of boating activities, estimated fuel use, and engine types was obtained from boat owner interviews at the Donner Lake public launching ramp. Boating activity surveys, conducted by binocular from three shore sites, provided estimates of the number of boats on the lake at a given time, together with their activity and general engine type. The level of detail obtained in the data set greatly reduced the number of assumptions inherent in the 1998 work and allowed calculation of reasonable estimates of MTBE loading for each day in the boating season.

Engine Emissions

Two of the project scientists from SB 521 have continued to study MTBE emission from marine engines since DLM-1998 was first released. Other members of the Lake Tahoe Motorized Watercraft Working Group[5] have also collected much better data on engine emissions through the use of tank tests (C. Emmett, CARB) and in-lake experimentation (G. Miller,

---

[5] The Lake Tahoe Motorized Watercraft Working Group was formed to address the issue of boating-related pollution in the lake. The group is composed of water-quality experts from agencies and universities, and representatives from the engine manufacturing industry. The group researched the issue and recommended a ban on certain types of two-stroke engines on Lake Tahoe.

University of Nevada, Reno). The 1998 data set differentiated only between two-strokes and non-two-strokes and only counted the emission contributions of the former. The new emission data allowed the marine engines encountered at Donner Lake to be subdivided into eight categories (see Table 3.10). Because the number of each type of engine was known, the MTBE contribution of each could then be calculated.

CHAPTER 3
RESEARCH FINDINGS

There are several research findings reported in this chapter. The results of a survey conducted among reservoir managers are presented as well as information collected on emissions from different types of engines. The results of work done at the Donner Lake test site are also presented.

The Reservoir Management Questionnaire (hereafter referred to as questionnaire) was sent to reservoir operators to collect information on reservoir characteristics, policies on motorized watercraft, concern over MTBE, water quality, and any MTBE monitoring and control strategies in place. The questionnaire found that, although concern over MTBE levels was prevalent, two-stroke engines were still common in many reservoirs.

A study of current marine engine technologies yielded data on emission rates that could be used in the model. Overall, conventional two-stroke engines released the greatest amount of fuel.

An extensive data collection program was instituted at Donner Lake in order to calibrate the model. A boating survey at Donner Lake (hereafter referred to as census) was conducted to determine use patterns for motorized watercraft, as well as fuel use by activity and engine type. Meteorological data were collected and sampling was conducted to monitor MTBE and BTEX levels in the lake.

RESERVOIR MANAGEMENT QUESTIONNAIRE

The questionnaire was conducted to collect information on adopted and proposed management options for reducing emission of MTBE and other fuel-related contaminants by various water utilities and water management agencies. The questionnaire was distributed to seventy-one water utilities / agencies in California, Texas, New Jersey, and New York. A list of agencies that responded to the questionnaire is provided in Appendix C.

Questionnaire Description and Objectives

The questionnaire (see Appendix D) was designed to obtain some basic facts about the subject reservoir, including its physical details, adopted or proposed policy on motorized watercraft, watercraft engine distribution, public and agency concerns about MTBE and other fuel-related contaminants, and adopted or proposed MTBE monitoring programs and control strategies.

The questionnaire contained the following four sections:

1. Contact Information: Questions in this section were used to obtain contact details of the utilities and the utility managers.
2. Basic Reservoir Information: This section focussed on obtaining information about the name and location of the reservoir(s) and whether motorized watercraft were allowed. This section also served to screen out those reservoirs that did not allow motorized watercraft. For reservoirs where these watercraft were permitted, information, such as the type of engines allowed, time periods of operation, and number of watercraft permitted, was collected. In addition, data on fueling facilities, MTBE content in the fuel, approximate watercraft engine distribution, and types of other activities permitted at the reservoir were also requested. In this section, respondents were also asked to indicate the presence and level of any public or governmental concern regarding MTBE and motorized watercraft.
3. Physical Details of the Reservoir: The purpose of this section was to obtain information on the hydrology of the reservoir. Information was collected regarding surface area, volume, annual range, source of water, land use around the lake/reservoir (to understand the watershed and possible sources of contamination), as well as the meteorological parameters monitored by the managing agency.
4. Water Quality Information: The purpose of this section was to obtain information on the existence of an MTBE monitoring program and the level of MTBE monitoring. Information was also being sought on the existence and extent of MTBE treatment in utilities where the reservoir was being used as a source of drinking water.

Results

Of the seventy-one questionnaires distributed, data were reported by forty-four reservoir managers (in California only) for a total of 89 reservoirs. The findings of the questionnaire presented in this section address the following topics: potential contamination of the reservoirs; the extent of public or agency concern over MTBE; extent of reservoir monitoring for MTBE and other parameters such as hydrology and meteorology that will be useful in understanding the fate and transport of MTBE and other fuel-related contaminants; and adopted or proposed strategies to reduce or prevent fuel-related contamination of the reservoirs.

Urbanization in Drainage Area

To understand the urbanization in the drainage area, the questionnaire requested information on the type of development immediately around the reservoir and in the drainage area. This information was used to understand the possible sources of urban pollution into the reservoir that could be a source of MTBE or influence its fate and transport. Of the 89 reservoirs for which data were provided by managers, only 36 percent provided information on urbanization in the drainage area. The following observations were drawn from the questionnaire results:

- The main forms of land use in the drainage areas around the reservoirs are broadly categorized as recreational, residential (urban and rural), grazing / agricultural, and restricted forest service lands.
- A high degree of urbanization, including temporary and permanent residences with supporting infrastructure (such as airports, roads, and domestic wastewater management systems); and recreational areas, such as marinas, camping facilities, and parks, were reported around eight reservoirs including Lake Tahoe in the North Tahoe Public Utility District, Scotts Flat and Rollins in the Nevada Irrigation District, Long Valley Reservoir, servicing the Los Angeles Department of Water and Power, and Big Bear Lake, which is in the Big Bear Municipal Water District. Water quality at all these reservoirs is extensively influenced from urban runoff.

- Among the respondents, industrial land use in the drainage area was reported only around Sweetwater Reservoir, which supplies the Sweetwater Authority in San Diego County. The east side of the reservoir reportedly is a combination of residential, industrial, and open space. However, this area of the reservoir is protected with an urban runoff diversion system. The north side of the reservoir, where land use is primarily residential, is also protected by the diversion system in the form of a series of channels and ponds that intercept the runoff and divert it around the reservoir.
- A lower level of urbanization, which includes small urban and rural communities (occupying 1 to 5 acres of land) with natural or minimal landscaping and temporary housing, is reported around four reservoirs. Agricultural and urban runoff will influence the water quality of these lakes. None of these lakes and reservoirs reported any form of protection.
- Boat-launch sites have been reported around Faucherie (primitive launch facility), Jackson Meadows, and Rollins and Scotts Flat, all of which are part of the Nevada Irrigation District. These sites indicate the potential for fuel contamination, and, therefore, MTBE and BTEX in the waters.
- Grazing is one of the predominant land uses reported around the Anderson, Calero, and Coyote reservoirs, which are part of the Santa Clara Valley Water District. A small residential development (300 homes) was reported adjacent to Anderson. Water quality in the reservoirs with respect to MTBE is not likely to be affected by the land use.
- Forest Service property constitutes one of the main land uses around five lakes.
- Lake Casitas in the Casitas Municipal Water District is the only lake that reported a watershed on federal land, which has a protected status. The shoreline area is used for recreation activities, such as camping, fishing, and special events.

Overall, respondents reported that urbanization has the potential to influence water quality (with respect to MTBE and BTEX) in the lakes and reservoirs for which data were reported. However, Sweetwater Reservoir, which is managed by the Sweetwater Authority San Diego County, was the only reservoir that reported any form of catchment protection to prevent MTBE contamination from runoff.

Public and Agency Concern over MTBE

The questionnaire results show that of the 89 reservoirs on which information was received, 40 responded to the question regarding public and agency concern over MTBE contamination. Twenty-five percent of the reservoirs responded that concern is high over MTBE contamination, 39 percent and 27 percent, respectively, responded that concern was medium and low. Categorizing the lakes and reservoirs based on the concern over MTBE, Table E.1 in Appendix E summarizes the questionnaire findings in terms of MTBE monitoring programs adopted policies with respect to MTBE management, and other controls to reduce, prevent, and/or minimize MTBE contamination. This summary demonstrates the level of concern as compared to extent and type of control strategies.

Following are some of the salient points that can be drawn from Table E.1 in Appendix E.

- Although high concern has been expressed by a number of utilities, two-stroke engines continue to dominate. Twenty-seven percent of the reservoirs that expressed high concern reported that only four-stroke engines operated in their waters. In Lake Tahoe, two-stroke engines are banned. In lakes and reservoirs where medium concern was reported, equal numbers of two-stroke and four-stroke engines operated. Lake Henshaw, among lakes and reservoirs reporting low concern, reported that only four-stroke engines operated in the lake.
- Personalized watercraft (PWCs) are not allowed in 63 percent, 18 percent, and 58 percent of the reservoirs that reported high, medium, and low concern, respectively, about MTBE.
- Except for Salinas Reservoir, all those with high concern have a regular MTBE monitoring program.
- Lake Henshaw and Shasta have very high development in the catchment, but report low concern over MTBE. On Lake Henshaw, however, it was reported that only four-stroke engines operate. Big Bear Lake also has a very high level of development, and MTBE has been detected in the drinking water; however, the lake does not have a routine monitoring program.

One of the most effective illustrations of MTBE management is the experience of the East Bay Municipal Utility District (EBMUD). On March 10, 1998, EBMUD began implementation of an action plan to control MTBE and other gasoline pollutants in EBMUD's drinking water reservoirs (Berger 2001). The action plan involved:

- Restricting the use of high emission outboard engines at San Pablo Reservoir effective January 1, 2000.
- Banning the use of motorboat engines at San Pablo Reservoir that discharge any fuel pollutant into the water, effective January 1, 2002.
- Monitoring water quality at San Pablo, Pardee and Camanche Reservoirs
- Informing the boating public and drinking water customers how EBMUD is addressing contamination of its drinking water by MTBE and other gasoline compounds.
- Participating in legislative and regulatory efforts to control contamination of drinking water supplies by MTBE and other gasoline components.

EBMUD also evaluated the applicability and potential effectiveness of, among other strategies, changing to fuel that does not contain MTBE for rental boats at San Pablo, Pardee and Camanche Reservoirs and controlling the number and type of outboard engines allowed and restricting the hours of motor boat recreation.

Of the three reservoirs, San Pablo and Pardee are drinking water sources while Camanche is a flood control reservoir. The focus of the action plan and follow up work has therefore been on San Pablo and Pardee.

Since 1998, the rental fleet at all three reservoirs has been changed to four-stroke engines. Concessionaires are required to use fuel without MTBE. This is monitored through regular inspections during which the fuel is tested for the presence of MTBE.

The 1999 results of the monitoring program showed that all MTBE levels detected in drinking water supplied from the three reservoirs were significantly less than anticipated regulatory standards. Coincident with reduced recreational boating, MTBE levels in September and October 1999 continued to decrease at all three reservoirs from previously reported levels. At San Pablo, monthly averages were below 1.2 ppb in 1999 compared to below 2.7 ppb in 1998.

Beginning in 2001, a lower detection limit was used for MTBE analysis (from 0.5 ppb to 0.09 ppb). Since 1997, average and maximum MTBE values have fallen by 50 percent at San Pablo. Through a large part of 2001, MTBE concentrations at San Pablo were around 0.13 ppb; however, in July 2001 the concentration fell below 0.09 ppb. Historic MTBE monitoring data for all three reservoirs from 1998 to 2001 are presented in Figure 3.1.

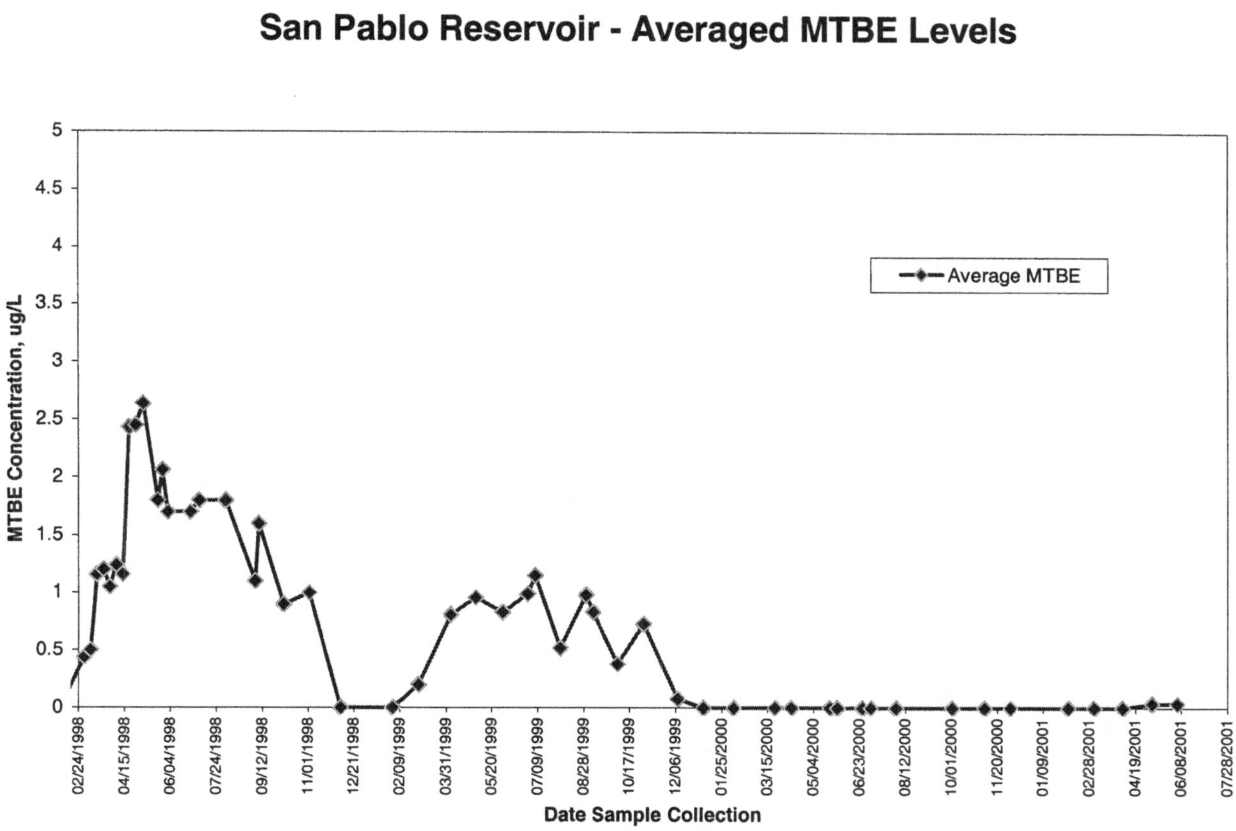

Figure 3.1 MTBE monitoring data for three California reservoirs (Figure courtesy of East Bay Municipal Utility District)

(continued)

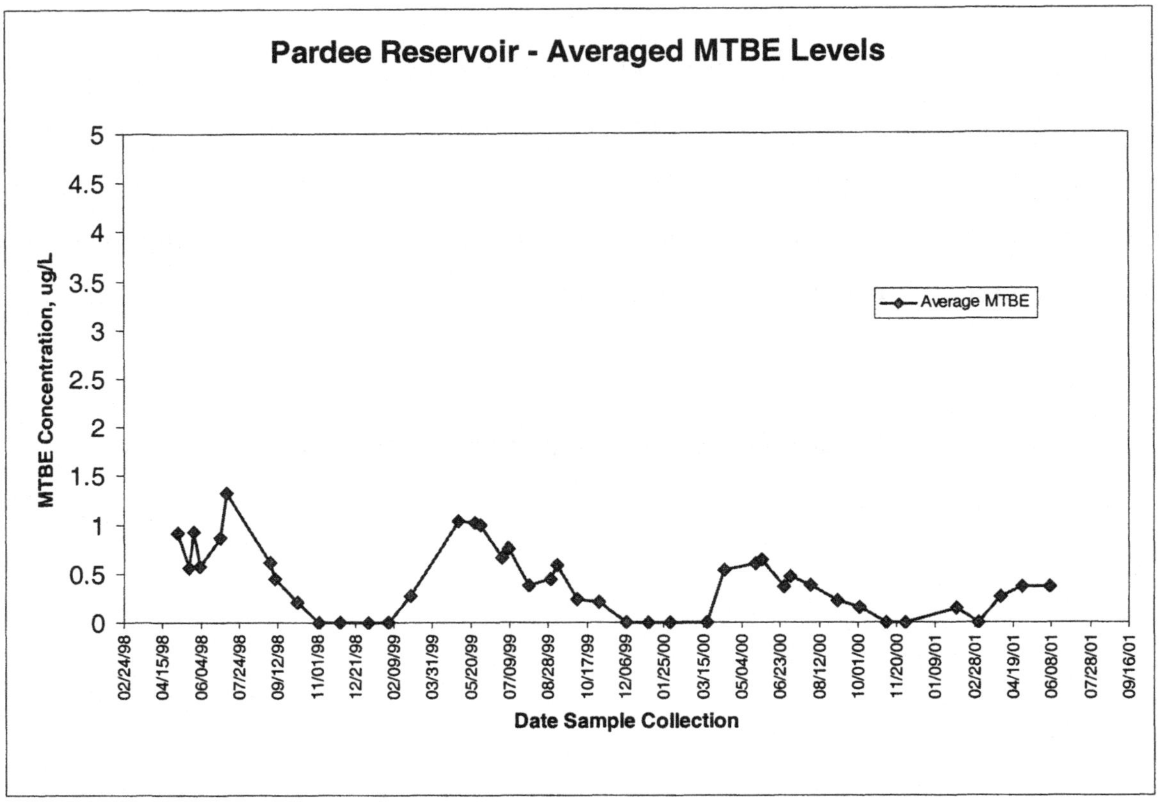

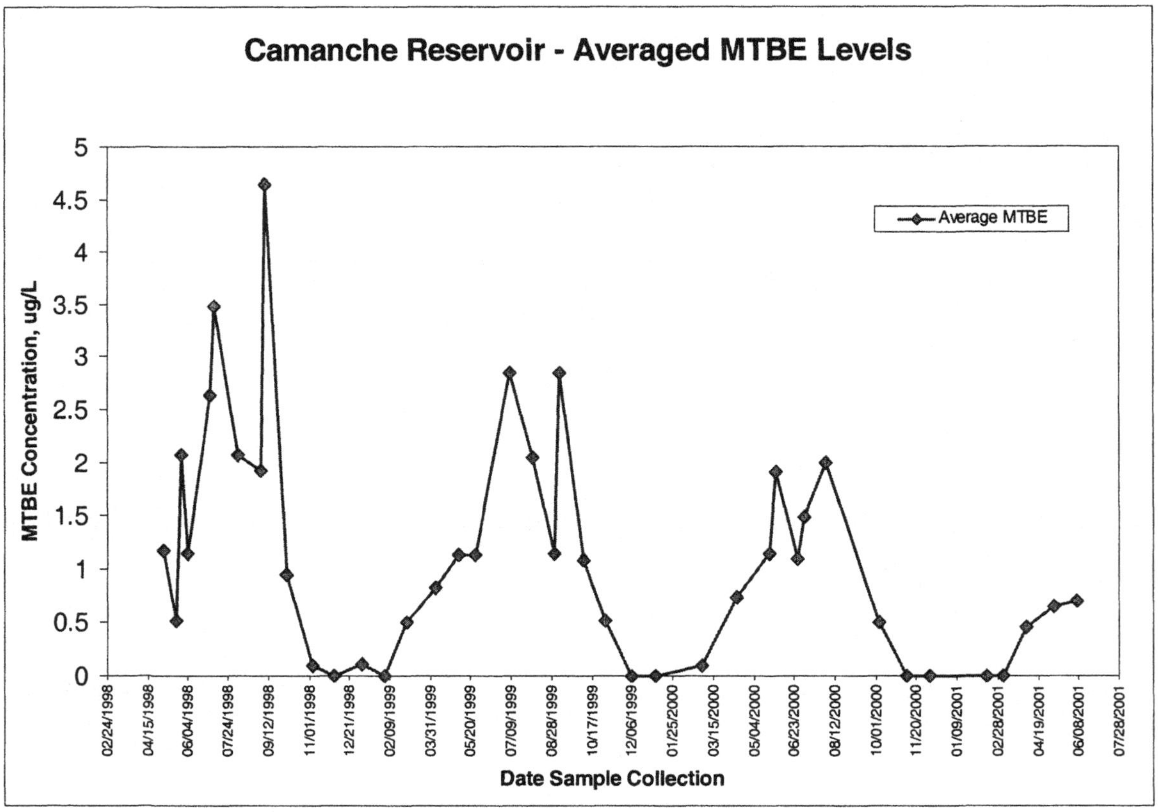

Figure 3.1 (continued)

As per the action plan, by 2002 all polluting outboard engines were to be banned effective January 2002. Only non-polluting engines (electric motors) were to be allowed on San Pablo. However, based on the MTBE data for 2001, which indicated that levels were below detection limits, the January 2002 ban on all polluting engines has been indefinitely postponed by EBMUD.

Reservoir Monitoring

Hydrology and meteorology. Hydrology of the reservoirs is based on the physical details data requested in the questionnaire. The questionnaire results showed that of the 89 reservoirs about which information was received, hydrology data were provided for only 47 reservoirs. The information included parameters such as surface area (in acres), volume (in acre-feet), annual range of volume, and maximum depth.

Information on the source of water into the reservoir was also requested. Of the 89 reservoirs about which information was received, source information was provided by 50 reservoirs. The main source of water appears to be local runoff. Therefore, the nature of development and urbanization in the catchment has a significant influence on water quality. Imported water and upstream reservoirs have also been reported as sources of water.

Among the 47 reservoirs that provided information on hydrology, meteorological data are collected only at 31 reservoirs. Parameters for which data collection was assessed in the questionnaire included wind speed, water temperature, relative humidity, air temperature, and solar radiation. Data on all the meteorological parameters listed in the questionnaire were collected only at two reservoirs: Lewiston and Sweetwater. The other 29 reservoirs collect data only for some of the parameters. Meteorological data are required for the DLM model.

General Water Quality

MTBE: Responses to the questionnaire reported that, of the 89 reservoirs for which responses were received, 48 percent collect water quality data for MTBE and 41 percent maintain regular MTBE monitoring programs. MTBE was detected in the finished drinking water supply in 29 percent of the 48 percent that collect water quality data for MTBE.

Reservoir Management Strategies

The questionnaire found that motorized watercraft were not allowed in 45 percent of the 87 lakes and reservoirs that responded to the questionnaire, and 83 percent of this total did not allow the crafts because of water quality concerns. Fifty-two percent of the reservoirs do not allow motorized watercraft, and 83 percent of that total do not allow motorized watercraft because of water quality concerns. Of the 34 reservoirs (out of 67 reservoirs) that allow motorized watercraft, 70 percent have some other form of control in place for their use.

In response to the questionnaire, 23 reservoirs provided additional details about reservoir management strategies that focus on control strategies for watercraft (see Table E.1 in Appendix E).

The control strategies include:

- Prescribing speed limits on watercraft.
- Restricting movement of watercraft in specific areas of the lakes and reservoirs.
- Legally banning PWCs or two-stroke engines from operating on the lake and reservoir.
- Setting up treatment facilities to control MTBE levels in the drinking water being supplied by the lake and reservoir.
- Stormwater control channels to prevent contamination of lake and reservoir.

The questionnaire also found that of the 48 percent of the reservoirs that collected water quality data for MTBE, 29 percent considered changing the water quality-treatment process to manage MTBE contamination. Thirty-five percent of the 41 percent that supported a regular monitoring program for MTBE also considered alternative treatment strategies.

Marine Engine Technologies

Information on marine engine technologies is important for two reasons. Firstly, engine efficiency data are used in the DLM-2002 model. The model requires input data that quantify the amount of fuel used by various marine engines that is exhausted unburned into the water during use. MTBE and BTEX are not combustion products; therefore, the model needs to "know" how many engines (and of what type) were active on a lake/reservoir, how much fuel was used by each engine type, and what portion of the total fuel used was exhausted directly into the water.

Second, information on the most current engine technology (i.e., fuel efficiency and exhaust properties) is needed to help guide reservoir managers' decisions on allowable engine options. Therefore, discussion in this section will also cover the production and availability of "cleaner" engines as well as fleet turnover rates.

Operating Principles

A variety of outboard engine technologies are currently in production. These technologies vary greatly in their ability to control raw fuel emissions as part of the exhaust. Raw fuel leaving the engine combustion chamber is the primary means of adding BTEX compounds and MTBE to surface waters. The operating principles of three general engine technologies and a description of how the raw fuel enters the exhaust system are presented in this section.

Conventional Two-Stroke Engines

Conventional two-stroke engines emit the greatest amount of unburned fuel to surface waters because of the design of the fuel loading and exhausting of the cylinder. As the piston moves down the cylinder, it passes the fuel port, allowing a pre-mixed fuel-oil solution to flow into the combustion chamber. At the same time, the piston passes the exhaust port, allowing the combustion gasses to escape into the exhaust system. Because both the fuel port and exhaust port are open at the same time, some of the fuel and oil mixture entering the combustion chamber is allowed to escape with the exhaust gasses. This results in raw fuel and oil emissions to the surface water.

There are several variations of this conventional engine technology. In most cases, the fuel-oil mixture is directed around the combustion chamber. This design achieves greater distribution and better atomization of the fuel, resulting in more complete combustion. Unfortunately, all of these variations have an inherent problem in that the fuel port and exhaust port are open at the same time, allowing raw fuel and oil to escape the combustion chamber.

Electronic Fuel Injection

In an attempt to increase the fuel economy of conventional two-stroke engines, the electronic fuel injection system was developed. This technology controls the amount of fuel entering the crankcase, where it is mixed with the lubricating oil. An electronic fuel injector delivers a measured amount of fuel, as determined by the need of the engine. The drawback to this technology is that it does not solve the problem of raw fuel and oil entering the exhaust. The fuel-oil delivery to the combustion chamber is similar to that of the conventional two-stroke engine described above. Both the fuel port and the exhaust port are open at the same time, allowing the passage of raw fuel and oil into the exhaust system and, subsequently, to the surface water.

Direct Injection

Direct injection engine technology dramatically reduced the amount of raw fuel that enters the surface water. This was achieved by changing the configuration of the fuel port and exhaust port within the cylinder and by injecting controlled amounts of fuel at specific times in the piston cycle. Unlike with the conventional two-stroke technology, fuel is only injected into the combustion chamber when the exhaust port is closed. This prevents the bypass of raw fuel to the surface water. Additionally, direct fuel injection technology separates the fuel from the lubricating oil reservoir. This allows for measured amounts of each to be delivered to the combustion chamber as needed by the engine. Direct injection technology results in about a 90 percent decrease in raw fuel emissions to surface water.

Electric Engines

The cleanest outboard engines on the water are electric engines. There is no fuel involved because they receive their power from on-board batteries. This results in zero fuel emissions to the surface water. The drawback to electric engines is that they have limited power output. Although they may serve well for fishing and slow pleasure cruising, it is not currently possible to engage in high-speed recreational activities, such as waterskiing and wakeboarding, with electric engine propulsion. It is also important to recognize that electric engines may not be practical on large lakes or those that routinely experience high winds. A certain level of power may be necessary to ensure safe boating conditions.

Four-stroke Engines

The four events that take place in a two-stroke engine also occur in a four-stroke engine, but a four-stroke engine dedicates a complete stroke, or cycle, to each action. Four-stroke engines benefit from mechanical valves that regulate intake and exhaust flow, and contribute to their excellent fuel efficiency.

As the piston travels downward from the top of the cylinder, a mechanical intake valve opens, and a mixture of air and fuel is drawn into the engine. As the piston reaches the bottom of the stroke, the intake valve closes to complete the intake stroke. The inertia of the flywheel and drive train components forces the piston upward to compress the air / fuel mixture trapped in the cylinder by the closed valves. As the piston reaches the top of the cylinder to complete the compression stroke, the fuel mixture is ignited, and that initiates the combustion stroke. The pressure created by the burning fuel pushes the piston downward. A four-stroke engine has no ports in the cylinder wall, however, so the combustion pressure pushes against the piston to the bottom of the stroke. Inertia then forces the piston upward again, but now the exhaust valve is open, and that forces all of the exhaust gases out of the cylinder. As the piston reaches the top of the exhaust stroke, the exhaust valve closes to complete the fourth stroke. Then the entire process is repeated. The four-stroke engine lends itself to more efficient running and less wastage of fuel than a two-stroke; the mechanical valves better manage fuel flow into, and out of, the engine.

Review of Existing Emission Studies

The literature on this topic is limited with regard to MTBE. The new interest in MTBE has lead to some additional published work such as Gabele and Pyle (2000), who found a 3 to 10 percent exhaust rate to the water for two-stroke and four-stroke 9.9 horsepower (Hp) outboard engines, respectively. Early and continuing work by Juttner and his colleagues in Switzerland (mid-1970s to mid-1990s) provides quantitative information on BTEX compounds. Table 3.1

presents the emission rates from various engines based on information available to February 2001.

Table 3.1

Engine emission data currently available - February 2001

Engine Type	Horse-power	MTBE	Benzene	Toluene	Ethyl-B	Information Source
		Soluble Fraction Of Constituent (%)				
PERSONAL WATERCRAFT (PWC)						
Two-stroke PWC	80	10	8	11	NA	G.C. Miller - Tahoe Study*
Two-stroke Sea-Doo	110	8.2	1.13	NA	NA	CARB Report 2001
Two-stroke EFI PWC	110	NA	NA	NA	NA	G.C. Miller - Tahoe Study
Four-stroke prototype	105	1.52	0.45	NA	NA	CARB Report 2001
OUTBOARDS, TWO-STROKE CARBURETED						
Two-stroke Suzuki	9.9	NA	NA	NA	NA	Juttner 1995
Two-stroke Mercury	9.9	24.4	8.03	NA	NA	CARB Report 2001
Two-stroke OB	15	NA	NA	NA	NA	G.C. Miller - Tahoe Study
Two-stroke Yamaha	20	NA	NA	NA	NA	Juttner 1995
Two-stroke Evinrude DI	90	1.9	0.6	NA	NA	CARB Report 2001
OUTBOARDS, TWO-STROKE DIRECT-INJECTED						
Two-stroke Johnson	90	11.1	1.08	NA	NA	CARB Report 2001
Two-stroke OB	150	NA	NA	NA	NA	G.C. Miller - Tahoe Study

(continued)

Table 3.1 (continued)

Engine Type	Horse Power	Soluble Fraction Of Constituent (%) MTBE	Benzene	Toluene	Ethyl-B	Information Source
Two-stroke OB Ficht	150	0.7	NA	NA	NA	G.C. Miller - Tahoe Study
OUTBOARDS, FOUR-STROKE						
Four-stroke Mercury	9.9	0.02	1.74	NA	NA	CARB Report 2001
Four-stroke Yamaha	9.9	NA	NA	NA	NA	Juttner 1995
Four-stroke Honda	90	0.06	0.14	NA	NA	CARB Report 2001
INBOARDS						
Four-stroke IB	150	0.2	0.4	0.3	NA	G.C. Miller - Tahoe Study
Four-stroke IB	170	NA	NA	NA	NA	G.C. Miller - Tahoe Study

NA – not available

* The G.C Miller-Tahoe Study can be found in the Reference section under Fiore et al. 1998.

Field and lab experiments conducted by Dr. Glenn Miller (Fiore et al, 1998) and CARB demonstrated that two-stroke carbureted engines had the largest amount of unburned fuel passing through the engine and into the water. For small two-stroke outboards, nearly 28 percent of the MTBE initially contained in the watercraft's fuel tank was deposited into the water during operation. Similarly, 10 percent of the MTBE in fuel used by two-stroke PWCs was loaded to the lake. Values for both categories of four-stroke engines were lower, at two to 3 percent and 0.2 percent for four-stroke outboards and four-stroke inboard / outboards and inboards, respectively. The results for benzene and toluene were similar in that the two-stroke engines were also found to be much less efficient. For the two-stroke engines tested, 8 to 9 percent of the benzene in the craft's fuel tank was loaded to the water after use; for the four-stroke engines tested, these values were much lower. The amount of benzene and toluene found in the water following operation of the four-stroke engines ranged from 0.3 to 1.8 percent. In current research, the boat-use data, and estimates of fuel consumption and engine efficiency data, were used to estimate MTBE and BTEX loading to Donner Lake from boating activity.

Evidence for Marine Engines as the Principal Source of MTBE Contamination in Surface Waters

The potential for emission from motorized watercraft engines as an important source of MTBE to surface waters was not widely recognized as recently as 1995 to 1996. This was, in part, due to the focus given to groundwater contamination (Zogorski et al. 1996). However, co-occurring studies during 1996 to 1998 by the Metropolitan Water District of Southern California (MWDSC) (six drinking water reservoirs in southern California) and the UCD – Tahoe Research Group (UCD-TRG) (two natural lakes in northern California, plus a statewide survey), clearly identified motorized recreational watercraft as the predominant source of MTBE in these lakes and reservoirs. At the same time, monitoring by EBMUD in the San Francisco Bay Area was also linking MTBE levels in drinking water reservoirs to motorized boating activity (M. Tikkanen, unpublished data, EBMUD).

Southern California Reservoir Study. Dale et al. (2000) recently published a review of the MWDSC findings from an extensive monitoring program on six reservoirs: Lake Havasu, Lake Mathews, Castaic Lake, Silverwood Lake, Lake Skinner, and Lake Perris. Lake Mathews does not allow motorized watercraft and, therefore, was considered a control. The detection of MTBE in surface waters of Lake Perris prompted monitoring at the other reservoirs.

Lake Perris supports a very high boating population. In fact, Dale et al. (2000) reported that typical summer weekend boat use was as high as that seen over the Memorial Day and Independence Day holidays. Surface water MTBE in Lake Perris ranged from 15 to 25 μg/L during the period of March to September with individual maximum values as high as 45 μg/L (Dale et al. 2000). This corresponds to the high boat season; the reduction of MTBE during November-February, when boating is much less common, strongly pointed to recreational watercraft as a primary MTBE source. Similar results were obtained for the other reservoirs that allow motorized recreational boating. For Lake Mathews, which does not allow boating, MTBE was never detected above the detection limit of 1.0 μg/L in 68 samples. Lake Skinner, which allows only fishing boats, contained minimal concentrations of MTBE in the range of 1 to 3.2 μg/L. These findings lead Dale et al. (2000) to conclude that the predominant mode of MTBE contamination in their study was the use of motorized recreational watercraft; with two-stroke engines as the major contributors.

Donner Lake and Lake Tahoe Studies. In the late 1990s, when MTBE received national attention, few published studies existed on the concentrations, sources, and fate of MTBE in surface waters. A notable exception was the UCD-TRG investigation of MTBE in Donner Lake, California, which is a multiple-use lake/reservoir in the Sierra Nevada (Reuter et al. 1998). The purpose of this study was to determine the (1) relative contribution of motorized watercraft as a source of MTBE, (2) its seasonal distribution, (3) loss from the water column, (4) extent of vertical transport, and (5) persistence between years. Over 500 individual samples were taken for MTBE analysis between March 1997 and January 1998. MTBE measurements were made at nine individual depths from surface to bottom (0-60 meters [m]) on 16 dates. Recreational boating was clearly the most important source of MTBE. Statistically, 86 percent of the change in MTBE was explained by variation in motorized watercraft use. Neither highway runoff nor precipitation contributed significantly. MTBE concentration ranged from <0.1 μg/L to 12 μg/L. Between July 1 and July 7, 1997, whole-lake MTBE content rose dramatically from 115 kilograms (kg) to 365 kg, as a direct result of motorized watercraft activity. By January 1998, levels declined to a minimum of 15 kg, suggesting little inter-annual persistence (see Figure 3.2).

The major loss of MTBE appeared to be volatilization at the air-water interface, characterized by two distinct periods. During the boating season, when input was still occurring, MTBE decline was 1.2 kg / day (193 day half-life). At the end of the boating season, MTBE loss increased to 8.1 kg / day (14 day half-life). These values are supported by data from Lake Perris (Dale et al. 2000) and by modeling (McCord and Schladow 1998). Thermal stratification acted to retard MTBE transport to greater depths. Upon turnover, MTBE was able to reach greater depths, but only temporarily. During summer boating activity, when MTBE concentrations are seasonally higher, the risk to drinking water supplies depends on the location of water intake.

UCD-TRG (Allen et al. 1998), U.S.G.S-Carson City, Nevada (Boughton and Lico 1998), and the University of Nevada, Reno (UNR) (Fiore et al. 1998) also conducted a monitoring program for MTBE in Lake Tahoe (CA-NV) between 1997 and 1998. Findings were very similar to the previous investigations; MTBE was only found during the boating season, and concentrations were elevated (up to 55 μg/L) in areas of heavy boat traffic.

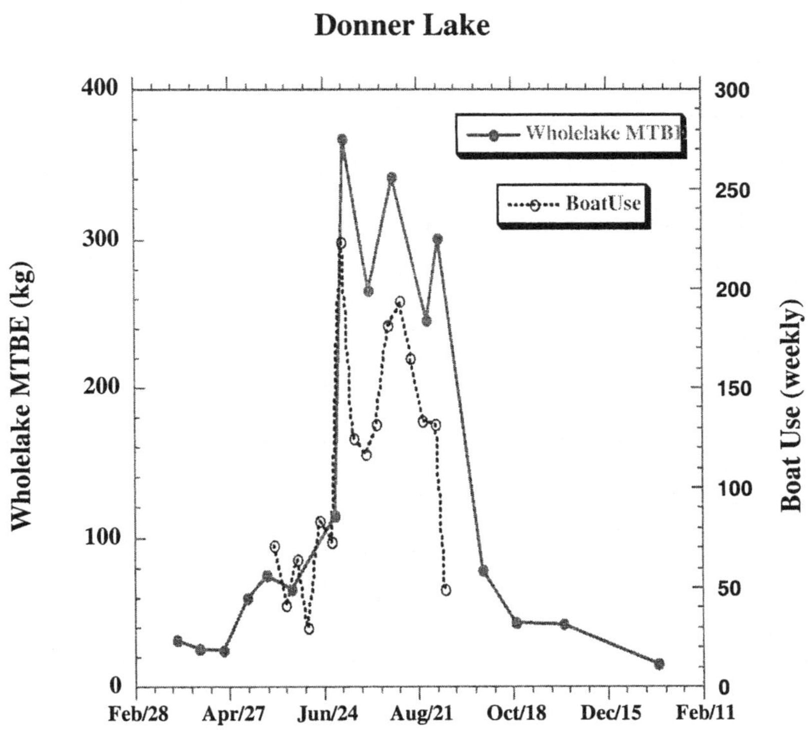

Figure 3.2 Effect of boat use on seasonal variation of MTBE concentration in Donner Lake

Survey of California Surface Drinking Water Supplies. At the request of the California legislature, the University of California conducted a health and environmental assessment of MTBE (Keller et al. 1998). As part of that study, Reuter et al. (1998) surveyed the available data on MTBE in the state's surface drinking water lakes and reservoirs between 1996 and 1997. This evaluation provided broad-based evidence that the connection between motorized boating and MTBE was not isolated to the few intensive studies discussed above; indeed, this appeared to be a statewide phenomenon.

Categorizing the 105 lakes and reservoirs for which MTBE data were available at that time, on the basis of MTBE detection and the allowed use of motorized watercraft, the data strongly support the hypothesis that the use of watercraft is an important source of MTBE to lakes and reservoirs. For the 59 lakes and reservoirs that allowed motorized watercraft, 44, or 75 percent, contained MTBE at concentrations above detection. Of the 15 lakes and reservoirs with boating but no detectable MTBE, seven were sampled in the winter when boating activity is significantly reduced. Viewed in a slightly different manner, 44, or 90 percent of the lakes and reservoirs with detectable MTBE allowed motorized watercraft. For the 46 lakes and reservoirs that excluded motorized boating, 41, or 89 percent did not contain MTBE at detectable levels. In the five lakes and reservoirs that contained MTBE at detectable concentrations, but which did not allow motorized boating, concentrations were low, ranging from approximately 0.6 to 2 μg/L.

Figure 3.3 provides summary data for the median concentration of MTBE in those lakes and reservoirs that allowed and excluded motorized watercraft. The data clearly show the strong relationship between MTBE concentration and motorized watercraft statewide. The median concentration in lakes and reservoirs that allow motorized watercraft was 3.6 μg/L, but only 0.2 to 0.3 μg/L where marine engines were not allowed. This lower value represents the limit of detection. Fifty percent of the concentrations in lakes and reservoirs with marine engines were in the range of 1.6 to 7.5 μg/L with values commonly in the 7.5 to 15 μg/L range. The maximum median level was 23 μg/L.

Summary. Much of the original published research linking MTBE contamination in surface lakes and reservoirs to motorized watercraft use was conducted in California, as described previously. Since then, investigators nationwide have confirmed these findings. For example, Baehr and Reilly (2001), reported that for a series of study lakes in northwestern New Jersey, the "source of the MTBE in the lakes was determined to be gasoline-powered watercraft." Other investigators have obtained similar findings; however, the goal of this section of the report is simply to establish the direct connection between marine engines and MTBE loading to surface waters. Based on these findings, an important approach for controlling MTBE in drinking water lakes and reservoirs is to manage boating. Of course, accidental spills can be very significant, should they occur in a specific lake/reservoir. For example, the accidental rupture of a petroleum pipeline near an inflow stream to Lake Tawakoni, Texas, resulted in concentrations of MTBE exceeding 1,000 ppb in parts of the reservoir during the first few days after the spill (Schladow, unpublished data). Lake Tawakoni is a water resource to the City of Dallas and this contamination resulted in the utility being unable to use this water until MTBE levels were below the health advisory value, i.e. <40 μg/L. While spill management is always important, it is far less common and much less predictable a source of contamination than emissions from motorized watercraft.

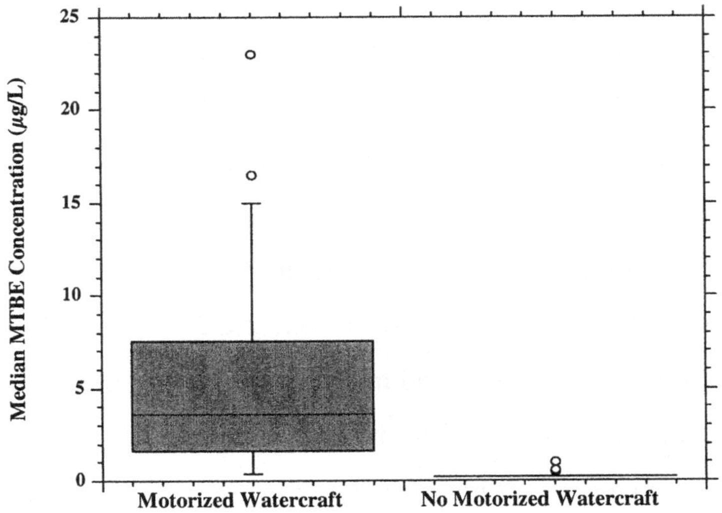

Figure 3.3 Summary data for the median concentration of MTBE in those lakes and reservoirs that allowed and excluded motorized watercraft. Whisker lines represent range of all data (with the exception of the identified outliers, denoted by the circle symbol). Each box encloses 50 percent of the data with the center line representing the median value.

Based on the conclusions above, the next step was to determine which type of marine engine was the largest contributor to MTBE loading.

Lake Tahoe Investigations

Background and approach. In response to the growing concern that the use of marine engines was having a negative impact on lake water quality, a scientific research team was convened to investigate the magnitude of fuel pollution that was entering Lake Tahoe from various types of watercraft. Participants in this Motorized Watercraft Technical Advisory Group included the UNR, UCD-TRG, USGS, Tahoe Regional Planning Agency, CARB, and the California Water Resources Control Board – Lahontan Region.

The total fuel load to Lake Tahoe from motorized watercraft emissions, and the relative contribution by various engine types have been calculated (Allen et al. 1998). These calculations use data from the 1998 boating survey (Hagler Bailly, Inc. 1998), laboratory emissions studies conducted by the CARB (CARB unpublished preliminary data), and in-lake enclosure studies conducted by UNR (UNR; Fiore et al. 1998). The contribution to fuel loading to Lake Tahoe from the various engine types requires knowledge of (1) the total amount of fuel used by each engine type during the boating season, (2) the percent of each water-quality constituent of concern in the fuel, and (3) the amount of each constituent deposited in the water by each engine type. The percent of fuel that passes through the engine and is deposited in the water upon exhaust is referred to as the released fraction. Calculations were done for MTBE and the BTEX compounds benzene and toluene. The discussion is largely limited to these two BTEX compounds because ethylbenzene and xylene(s) were only infrequently detected in the lake water and speculation was not appropriate.

The total amount of fuel use estimated at Lake Tahoe during the 1998 boating season was approximately 1.5 million gallons for a combined 486,500 hours of boating activity (Hagler Bailly, Inc. 1998). Classifying the gasoline-powered watercraft into four groups, the following estimate of fuel use was obtained: PWC - 96,793 gallons; two-stroke outboard - 79,504 gallons; four-stroke outboard - 20,524 gallons; and four-stroke other - 1,298,867. The two-stroke engines, including PWCs, used only 11 to 12 percent of the reported total gasoline consumption. All engines in the PWC category were assumed to use two-stroke carbureted technology. The two-stroke outboard category included all carbureted outboard marine engines of this type, regardless of Hp rating. The largest category of engine used during the 1998 boating season was the non-outboard, four-stroke engine. These engines included gasoline-powered inboard / outboards and inboards, which are typically found in ski boats and cruising pleasure craft. It is not surprising that this category used the largest volume of fuel because these engines made up over 50 percent of the total boat fleet on Lake Tahoe during the 1998 boating season.

The fraction of gasoline represented by MTBE, benzene, and toluene was 14 percent, two percent and 8 percent, respectively. This is the percent of each constituent found in California Certified Fuel by weight, not volume. For example, MTBE is present as 11 percent by volume.

The soluble or released fraction of MTBE, benzene, and toluene were calculated from data provided by the CARB and the UNR studies. The CARB laboratory studies were primarily intended to compare emissions from different marine engines with concurrent water-quality and air-quality monitoring for each engine test. These investigations used a 9,842-liter experimental tank. The UNR study allowed them to take measurements directly in the field from an isolated area of Lake Tahoe, where the study team operated watercraft with various engines as part of a designed experiment. Both studies provide the weight of fuel that was consumed by the engine during test operation, the weight equivalent of MTBE and BTEX compounds in the fuel, and the concentrations of these compounds in the water after each engine test. From these data, the soluble fractions were calculated for each engine type. The soluble fractions are different for each engine tested and for each of the fuel constituents. For field experiments, all water samples for the MTBE/BTEX analyses were collected within <1 hour after the test (i.e. termination of engine operation). The CARB experiments allowed for sampled collection at a number of time intervals following termination of engine use (CARB 2001). To calculate engine efficiency, samples taken at time-zero (i.e., immediately after the engine test) were used.

The amount of fuel constituent deposited in the water by each engine type was calculated as follows: the soluble fraction of fuel constituents for PWCs was calculated from in-lake enclosure studies using an 80 Hp carbureted two-stroke (Fiore et al. 1998). Fuel load attributed to the two-stroke outboard engines was based on the CARB tests (Mercury 9.9 Hp two-stroke) which provided data on fuel used per test, along with the resulting concentration in the experimental tank. The four-stroke outboard fuel loads were also calculated from CARB data, as above, and using a Mercury 9.9 Hp four-stroke engine. The soluble fraction attributed to the four-stroke - "other" category was derived from in-lake enclosure studies conducted by UNR (170 Hp four-stroke inboard / outboard engine). It should be noted that the engines used in these tests merely serve as surrogates for the class of engine within the fleet. For this reason, calculations of fuel load were considered estimates.

California – CARB Tank Study (2001). Since the Lake Tahoe study was first published in 1998, CARB published a final report on their marine engine emissions study (CARB 2001). This recent document contains results from additional engines that were not available for the Lake Tahoe analysis.

Table 3.2 summarizes the results of MTBE, BTEX, and benzene emission data from CARB (2001) and Fiore et al. (1998). Fiore et al. (1998) took water samples during field experiments in Lake Tahoe, and typically collected these samples in 1 hour. A lakeside enclosure with the dimensions of 100 by 200 feet, with an average depth of 6 feet, was used. Watercraft were operated for 30 minutes on sequential days. The PWC ran at operational speeds; however, the larger engines could only run at trolling speeds. This could have over-estimated fuel efficiency and underestimated fuel release for these engines (Fiore et al. 1998).

Table 3.2

Summary of the results of MTBE, BTEX, and benzene emission data from CARB (2001) and Fiore et al. (1998)[1]

Engine Type	MTBE	BTEX	Benzene	Ref.
Two-stroke carbureted PWC - 80 Hp	10	ND	8	Fiore et al. (1998)
Two-stroke carbureted PWC – 110 Hp	8.2	1.2	1.1	CARB (2001)
Two-stroke carbureted outboard - 9.9 Hp	27.9	11.0	8.0	CARB (2001)
Two-stroke carbureted outboard – 90 Hp	11.1	1.1	1.1	CARB (2001)
Two-stroke direct-fuel injected outboard - 90 Hp	1.9	0.3	0.6	CARB (2001)
Two-stroke direct-fuel injected outboard/inboard – 150 (Ficht)	0.7	ND	0.5	Fiore et al. (1998)
Four-stroke direct-fuel injected PWC – 105 Hp	1.5	0.2	0.5	Fiore et al. (1998)
Four-stroke carbureted outboard - 9.9 Hp	2.0	1.4	1.7	CARB (2001)
Four-stroke carbureted outboard – 90 Hp	0.06	0.05	0.14	CARB (2001)
Four-stroke carbureted outboard/inboard – 170 Hp	0.2	ND	0.4	Fiore et al. (1998)

[1] Values represent percent of fuel constituent that was released in water following operation of the listed marine engines, based on field and laboratory testing (i.e., emission rate).
ND – non detect

CARB (2001) used a laboratory tank (9,842 liters) especially designed to sample the aqueous and air components following engine operation. CARB's water samples were taken immediately after the experiment ended at a nominal time (zero). The test cycle consisted of a two-minute warm-up, followed by five different throttle positions, each lasting for seven minutes. A series of four accelerations and decelerations were included for a total engine operation period of 60 minutes (CARB 2001). Efficiency values were calculated as the ratio of amount of fuel constituent found in the tank water relative to reported fuel consumption.

The following generalizations are relevant to our current modeling and management options studies. As anticipated, aqueous emissions of MTBE from two-stroke carbureted engines (8.2 percent to 27.9 percent) were considerably higher than from four-stroke carbureted engines (0.06 percent to 2.0 percent). The mean two-stroke: four-stroke ratio for all the reported carbureted engines was 19, and in reasonable agreement, albeit slightly higher, with that reported by Gabele and Pyle (2000). CARB's test of a two-stroke carbureted (9.9 Hp Mercury) engine and a four-stroke carbureted (9.9 Hp Mercury) engine allowed for a direct comparison, yielding a ratio of 14.0 for MTBE. In all cases, when comparable engines were tested, reduced fuel Hp resulted in lower efficiency and increased release. For example, the 9.9 Hp two-stroke carbureted outboard engine released 27.9 percent of its used fuel to the water, compared to 11.1 percent for a 90 Hp engine.

Two-stroke, direct-fuel-injected engines behaved most similarly with the four-stroke in terms of MTBE release. As suggested with the Lake Tahoe data above, the use of four-stroke and two-stroke, direct-fuel-injected marine engines provide technological alternatives to the standard two-stroke carbureted engines. For example, there was an 80 to 85 percent reduction in MTBE release from PWCs when using the four-stroke, fuel-injected, 105 Hp engine compared to a two-stroke, carbureted, 110 Hp engine.

In general, BTEX and benzene followed a similar pattern, although the relationship to MTBE was much less clear. BTEX and benzene released from the 2SC-PWC-110 Hp and 2SC-OB-90 Hp engines used by CARB were less than expected.

Based on the volumes of fuel used by each class of engine, as reported by the watercraft use study and the calculated soluble fractions, carbureted two-stroke engines contributed a disproportionate load of MTBE, benzene, and toluene to Lake Tahoe. If the PWC and two-stroke outboard classes are combined to represent the two-stroke carbureted engines, they cumulatively contributed over 90 percent of the MTBE to Lake Tahoe, while only using about 11 to 12 percent of the total fuel used by boats in 1998. Moreover, these engines were responsible for over 70 percent of the benzene and 80 percent of the toluene deposited during the boating season. In contrast to this, the four-stroke inboard and inboard/outboard class consumed 87 percent of the fuel used by boating on Tahoe, but was responsible for 7 percent of the estimated MTBE, 28 percent of the estimated benzene, and 17 percent of the estimate toluene loading to the lake (Figures 3.4 and 3.5). Expressed as gallons of constituent loaded, two-stroke engines contributed thousands of gallons of MTBE, hundreds of gallons of benzene, and tens of hundreds of gallons of toluene.

The soluble fraction of fuel constituents from a Ficht (direct) Injected two-stroke engine tested in Lake Tahoe water was found by Fiore et al. (1998) to be much lower than the traditional carbureted two-stroke engines. For all three of the water-quality constituents (MTBE, benzene, and toluene), the percent of these compounds loaded to the water from fuel in the gas tank was on the order of 0.5 percent. Calculations were done to assess what changes might have been realized in the 1998 fuel constituent load to Lake Tahoe if all the carbureted two-stroke engines had soluble fractions equivalent to the Ficht Injected test engine. Using the same volumes of fuel reported in the watercraft-use study, nearly 90 percent of the MTBE, 70 percent of the benzene, and 80 percent of the toluene could have been kept out of the lake. These savings would be even more dramatic if appropriate fuel use numbers for these engines were available. It is estimated that fuel use could be reduced by as much as 25 percent using the more efficient injected two-stroke technology.

Although these calculations incorporate the stated assumptions and the fuel constituent loads are estimates, it is clear that the carbureted two-stroke engines add a disproportionate amount of the fuel constituents to Lake Tahoe. It is also clear that the newer technology-injected two-stroke engines can greatly reduce fuel loads to the lake.

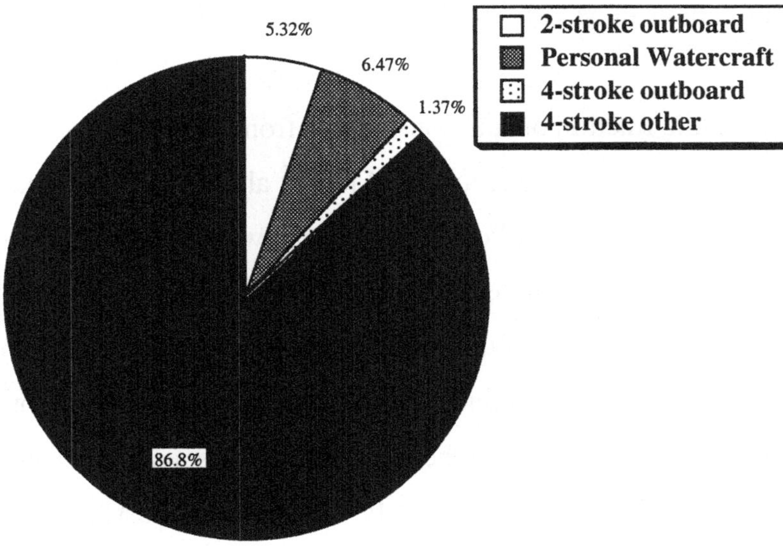

Figure 3.4 Fuel use by engine

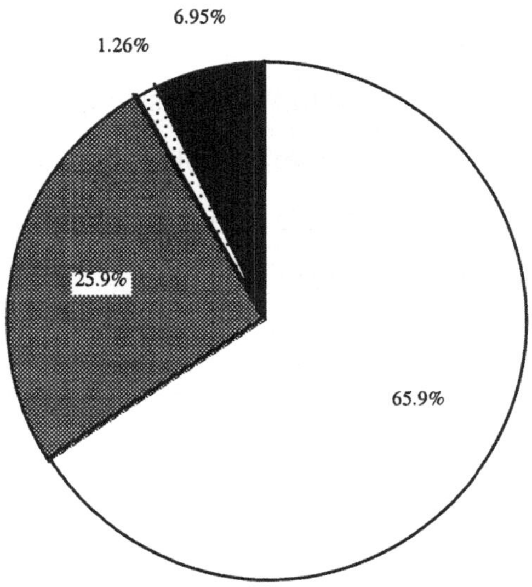

Figure 3.5 Estimated MTBE load by engine type

Effect of two-stroke ban: On the basis of data from the Lake Tahoe study and other considerations, certain two-stroke engine technologies were banned at Lake Tahoe. Following this ban, ambient concentrations in many locations around the lake decreased by a factor of 10.

Effective June 1, 1999, the Tahoe Regional Planning Agency imposed a ban on most carbureted two-stroke marine engines. Lake Tahoe was monitored in late August and over the Labor Day weekend in 1999 to determine if concentrations of MTBE and BTEX were affected by implementation of the new marine engine policy (Allen and Reuter 1999). Samples were taken on three dates: mid-week, weekend, and following Labor Day; peak boating activity traditionally occurs at this time of the year. Samples were taken from mid-lake (open-water locations), ten stations around the perimeter of the lake, and ten "hot spots" along the South Shore, where boating and boating-related activities are high.

The results showed a significant decline in both MTBE and BTEX compounds between 1997 and 1999. Generally, ambient concentrations of MTBE decreased by an order of magnitude (a factor of 10), with samples around the north end of the lake at or below the 0.06 μg/L level of detection. Samples collected near the south end of the lake showed a similar drop in concentration from previous years, but remained above the level of detection at a few tenths of a part per billion (μg/L) (Figure 3.6).

High levels of MTBE were still found at certain hot spots, which were attributed to isolated boating actions not related to overall boating intensity (draining of bilge water when trailering out a boat, spillage during refueling, etc.). Ambient concentrations of the BTEX compounds (benzene, toluene, ethylbenzene, and xylene) at the nearshore locations were also found to be lower than levels recorded in the past two years of monitoring. For example, Allen and Reuter (1999) reported an 88 percent reduction in toluene after the ban in 1999.

Combined, the data strongly suggest that the ban on certain types of two-stroke engines at Lake Tahoe was very successful in reducing both MTBE and BTEX.

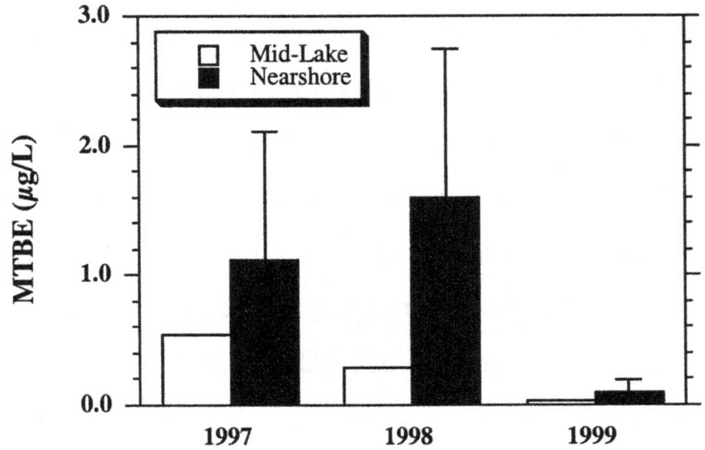

Figure 3.6 Effect of the two-stroke ban at Lake Tahoe (Error bars represent one standard deviation.)

Application of Marine Engine Technologies Study to the Project

The UCD DLM-2002 Model requires input data characterizing the release of fuel constituents into the water during marine engine use. Using the studies and data presented above, the authors of this study selected emission fractions as shown in Table 3.3. The listed engine types represent the major categories that were observed during the 2000 census at Donner Lake, and those which are represented in the model. Since the field measurements for BTEX compounds in Donner Lake during the 2000 boating season were so infrequent and low, this study focused on MTBE.

Table 3.3

Emission fractions of MTBE from marine engines into lake water during use

Engine Type	Emission Fraction	Notes
PWC, two-stroke, carbureted	0.09	Mean of 80 and 110 Hp two-stroke carbureted PWC engines
PWC, two-stroke, direct injected	0.03	Determined by comparing MTBE released during single pass experiments with two-stroke PWC (80 Hp) and two-stroke electronic fuel injected PWC (110 Hp) (Fiore et al. 1998)

(continued)

Table 3.3 (continued)

Engine Type	Emission Fraction	Notes
Outboard, two-stroke, carbureted □25 hp	0.28	Two-stroke carbureted outboard (9.9 Hp) engine
Outboard, two-stroke, carbureted >25 hp □75 hp	0.20	Mean of two-stroke carbureted outboard (9.9 Hp) and two-stroke carbureted outboard (90 Hp) engines
Outboard, two-stroke, carbureted >75 hp	0.11	Two-stroke carbureted outboard (90 Hp) engine
Outboard, two-stroke, direct injected	0.01	Mean of two-stroke direct fuel injection outboard (90 Hp) and two-stroke direct fuel injection inboard/outboard (Ficht)(150 Hp) engines
Outboard, four-stroke	0.01	Mean of four-stroke carbureted outboard (9.9 Hp) and four-stroke carbureted outboard (90 Hp) engines
Inboard, including stern drive	0.002	Four-stroke carbureted inboard / outboard (170 Hp) engine

(These values were used in the UCD DLM-2002 model. A value of 1.0 means that 100 percent of the MTBE in the fuel passed unburned into the receiving reservoir).

Selection of emission rates for model. Review of literature on efficiencies of various marine engines indicated that while much of the literature deals with specific compounds produced during engine operation, it does not address the volume of the constituents loaded to the water based on the volume of fuel consumed by the engine. This missing information does not allow for the calculation of engine efficiency. Therefore, the primary sources of soluble fraction information are from Dr. Glen Miller at CARB, and Mr. Gabele and Mr. Pyle with USEPA. Together, they cover the efficiencies of 12 engines, including clean technology and conventional carbureted PWCs and outboards. The soluble fractions of fuel constituents determined by these studies will form the basis for input into DLM-2002.

The USEPA and California emission standards. The USEPA adopted atmospheric emissions standards for marine engines that are designed to meet the requirements of the Clean Air Act. These standards have become known as the "EPA 2006 standards." They require a 75 percent reduction in emissions of hydrocarbons and oxides of nitrogen to the air by 2006. However, the standards do not apply to individual engines, the 75 percent reduction is for the entire fleet of engines that an individual manufacturer produces. Therefore, a company could continue to produce conventional two-stroke engines as long as they are producing enough clean technology engines (direct-injected two-stroke or four-stroke) such that the 75-percent reduction is met for their entire product line. Further confusion arises when an individual engine is said to meet the USEPA 2006 standards. This claim is stating that the specific engine has been shown to have at least a 75 percent reduction in atmospheric hydrocarbons and nitrogen oxides over the USEPA baseline for that engine's power output. Many engines are now being marked with stickers stating that they meet the USEPA 2006 standard.

Two questions still remain: (1) What does this air standard mean for water quality, and (2) will the EPA 2006 standards prevent discharges of fuel to lakes and reservoirs from marine engines after its effective date?

While the USEPA 2006 standards regulate air emissions from marine engines, a secondary, albeit important contribution is the dramatic reduction in the amount of fuel entering the water. To meet the reduction in emitted hydrocarbons and nitrogen oxides to the air, manufacturers had to eliminate the "blow by" that is inherent in conventional two-stroke engines. "Blow by" occurs when both the fuel port and exhaust port in the cylinder are open at the same time. Some of the fuel entering the cylinder passes through to the exhaust and leaves the engine as uncombusted fuel. New two-stroke technology that meets the USEPA 2006 standards uses some form of direct fuel injection. The fuel is injected under pressure to the cylinder only after the exhaust port has been closed. This allows for more complete combustion of the fuel and eliminates large discharges of raw fuel to the exhaust system. Other forms of technology have been tried but have not been able to meet the USEPA standards. These other technologies include injecting the fuel into the engine's crankcase but not changing the inflow-outflow design of the fuel in the cylinder. Catalysts were also added to the exhaust systems in an attempt to combust fuel that has escaped unburned through the "blow by" process. While these technologies have shown some reduction in emissions, they have not improved the fuel efficiency of the engines.

Unfortunately, the USEPA 2006 standards do not guarantee that all engines being used after 2006 will have reduced emissions. As described above, the standards apply to a "fleet average," and engine manufacturers can continue to produce conventional two-stroke engines for the marketplace. It is entirely possible that an individual lake/reservoir might not experience any change in engine technology as a result of the USEPA regulations. The only way to ensure that more efficient engines are operating on a given lake/reservoir is through local management.

Since water managers will be responsible for deciding which types of engine technologies will be allowed on specific lakes and reservoirs, it is important to know which marine engine products are currently available.

Overview of 2001 engine technology. PWC: Most of the major makers of PWCs have recognized the need to incorporate new, cleaner engine technology into their fleet of boats. This was probably the result of strict regulations designed to protect water quality, which limited or banned conventional two-stroke technology from certain lakes and reservoirs (i.e., Lake Tahoe). Both Bombardier (Sea-Doo) and Kawasaki (Jet Ski) are selling boats in 2001 that meet the USEPA 2006 standards. The number of models that are equipped with the clean technology is limited. Both Sea-Doo and Kawasaki are including the technology in one two-passenger and one three-passenger vessel. In both cases, the PWC only comes in the highest available Hp, which is 130 Hp, making it impractical to be used in most recreational rental fleets. PWC rental concessionaires are often reluctant to provide high-performance PWCs to novice riders.

Honda brought the first four-stroke PWCs to market in 2002. It meets USEPA 2006 standards. However, Honda has yet to release information on the Hp, number of passengers, and number of models available with the new technology. Yamaha, the maker of Wave Runners, is not selling any craft in 2001 that meets the USEPA 2006 standards.

The future expansion of clean-engine technology (both direct-injected two-stroke and four-stroke) in the PWC market will likely depend on consumer demand. The major manufacturers have met their obligation to produce clean-engine alternatives. Consumers will determine if these engines become the standard for all models across the Hp range. More details on various models of PWCs are presented in Appendix F.

Outboard Engines: The outboard engine manufacturers have expanded the options for clean-engine consumers. Each of the major brand names has engines available that span the range from almost 2 Hp to 250 Hp, and meet USEPA 2006 standards. Generally, the low-emission engines that are less than 100 Hp are represented by four-stroke technology, while those greater than 100 Hp are some type of injected two-stroke technology. It appears likely that there will be a reduction in U.S. sales and production of conventional two-stroke engines prior to 2006. There does not appear to be a large discrepancy in the price of the two technologies (clean vs. conventional), and the new cleaner engines have been around long enough to have proven their reliability. This projection is supported by the fact that Mercury Marine will no longer sell conventional two-stroke engines in the U.S. or Canada after 2006. It is unlikely a major producer of outboard motors would make this move if they felt they would loose market share as a result.

The greatest turnover in outboard motors will most likely come from rental fleets. These businesses are regulated very closely because they are seen as the major source of engine hours on many lakes and reservoirs. Additionally, a concessionaire would see the economic gain from fuel savings as offsetting the cost of purchasing new engines.

Outboard engines are typically grouped into four categories based on the Hp rating of the engines. Portables generally range from the smallest gas powered motors (about 2 Hp) up to 8 Hp and are often used as auxiliary trolling motors for fishing. The cost differential between conventional two-strokes and those meeting the USEPA 2006 standards is between $200 and $400. However it is important to note that some four-stroke engines in this category are cheaper than conventional two-strokes of the same Hp.

Compact engines range from 9.9 to 25 hp. These are usually considered trolling motors, as well, but they have enough power to propel a small boat to and from the fishing location. Engines with clean technology in this Hp range cost between $200 and $800 more than their equivalent conventional two-stroke counterparts.

The next larger size category is the mid-range engines. These are generally between 40 and 115 Hp. The engines in this category make the shift from small trolling motors to those that are used for other forms of recreation (tubing, wake boarding, water skiing). It is also in this category where the shift in clean engine technology changes from four-stroke engines to direct-injected engines. The largest four-stroke, outboard engine currently available is the Honda 130 Hp. The cost difference between the engine technologies jumps dramatically in this range.

Consumers could expect to pay from $1,500 to $2,800 more for a cleaner engine of comparable power.

The largest engine category of outboard motors ranges from 130 to 250 Hp and is typically used for larger boats and recreation requiring higher speeds. Various types of direct-injected technology represent all clean engines available in this category. As a result of the size of the engines and the very new technologies being marketed, the cost differential between engines meeting USEPA 2006 standards and those that don't is quite high. On average, a clean engine will cost nearly $2,700 more in this category. More details on outboard engine models are presented in Appendix F.

It is important to note that the difference in price between technologies is for comparison purposes only, and is based on reported manufacturer's suggested retail price. It may be possible for consumers to purchase engines at significantly reduced prices at other venues.

DONNER LAKE TEST SITE

Donner Lake is located in the Sierra Nevada, just east of Donner Pass and south of I-80. It is positioned approximately 38 km west of Reno, Nevada, and 20 km northwest of Lake Tahoe (see Figure 3.7, which also shows the locations of the boating census and water quality monitoring program stations; these activities are discussed later in this chapter). The majority of inflow water enters the lake as snowmelt. A concrete weir structure on the outlet stream provides approximately 3 meters (m) of controlled water storage. Undiverted water subsequently flows into the lower Truckee River. Small volumes of water are also withdrawn from approximately 10-m depth by the Donner Lake Water Company for local, domestic service. Recreational activity is largely confined to the summer months, between Memorial Day and Labor Day because of the cold winter climate and the distance from population centers. Typical boat use is trolling fish boats (6 a.m. to 10 a.m.), ski boats (10 a.m. to 2 p.m.), and jet skis (10 a.m. to 6 p.m.). Approximately 60 percent of the boats are powered by two-stroke engines (Allen, unpublished data).

Figure 3.7 Census and water quality monitoring program locations at Donner Lake, CA

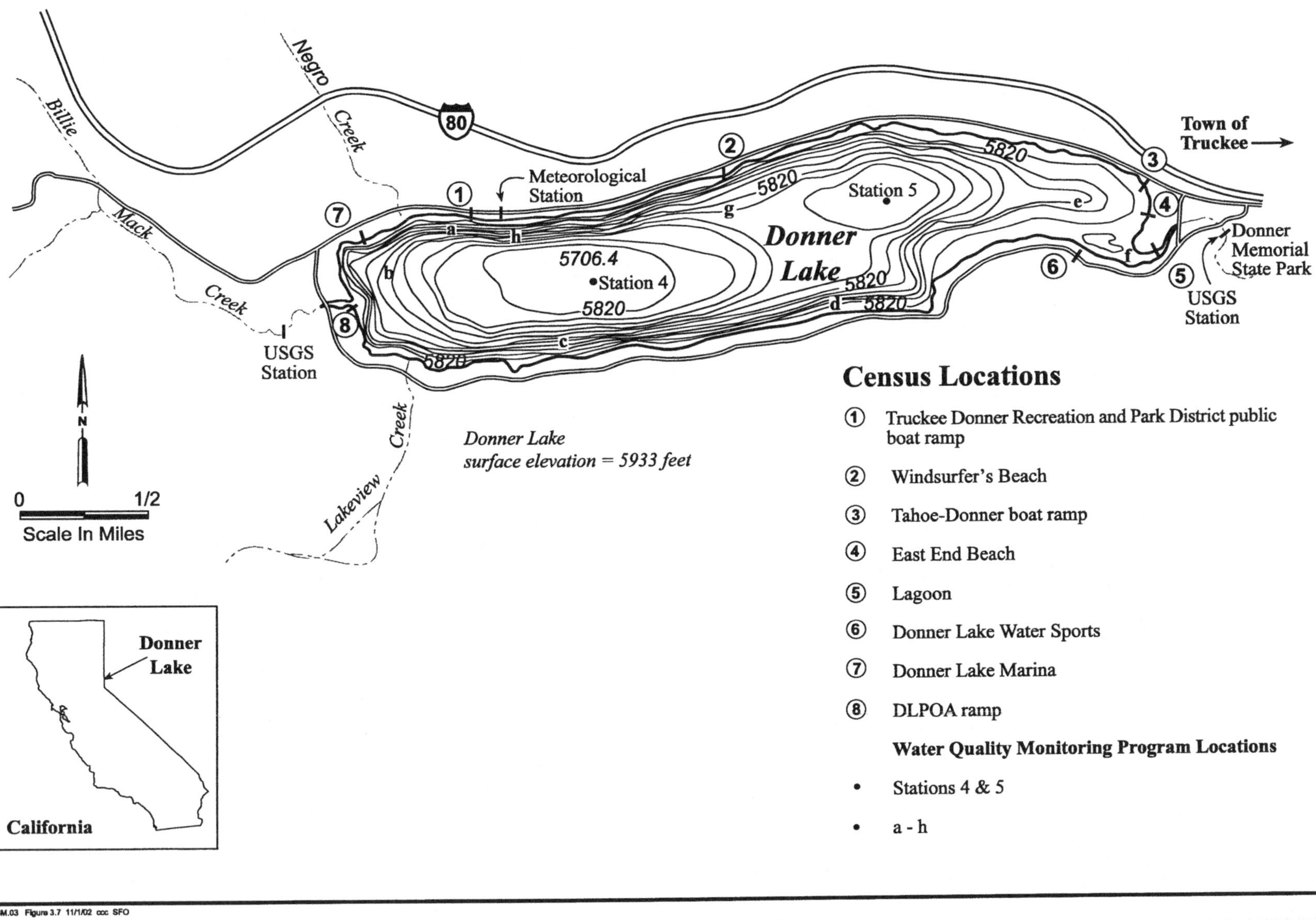

Physical Characteristics of the Lake

Lake Characteristics and Hydrology

Donner Lake lies at an elevation of 1,809 m above sea level and occupies a tectonic basin. With a surface area of 3.9 square kilometers (km^2) and a drainage basin of 36 km^2, the watershed-to-lake area ratio is 9.2. The lake has a two m dam at its outflow, which provides a reservoir pool; however, most of the lake volume (approximately 1.26 x 10^8 m^3, or 102,000 acre-feet) is below this reservoir pool (±92 percent). Donner Lake has a maximum depth of 70 m and an average depth of 33 m. The ratio of annual inflow to total lake volume is on the order of 1:10.

Spring snowmelt, which typically occurs between April and June, is the dominant input feature in the water budget. Summers are extremely dry in the Sierra Nevada, with the exception of occasional, brief thunderstorms that do not contribute significantly to the water budget of the medium-to-large subalpine lakes. Lake elevation data collected by the U.S.G.S in 2000 showed an increase in lake level throughout May due to snowmelt; this rise peaked on May 25 at an elevation of 1809.77 m. Elevation remained very stable, at or about 1809.65 m, until June 22, indicating that inflow was in equilibrium with outflow and evaporative losses. Lake level declined steadily during the summer months of July through September at a rate of approximately 0.8 centimeters per day (cm/d) as a result of evaporation and no significant precipitation. Between September 21 and October 31, lake level dropped at an increased rate of 3.5 cm/d. This increased outflow is a regular feature of the lake's hydrology and is part of a managed release schedule. The schedule provides increased water supply for downstream use at the end of the dry summer, and allows additional holding capacity in Donner Lake to accept fall rainstorms, should they occur.

Among its designated beneficial uses, Donner Lake is a source of drinking water for lakeside residents, and contributes to the water supply for downstream communities including Reno, Nevada. The sole outlet is Donner Creek, which drains directly into the Truckee River. Lake descriptions were taken from the following sources: Dong (1974), Vineyard (1994), and Reuter et al. (1998).

Limnology

Donner Lake has not been the subject of extensive or long-term studies of its physical, chemical and biological characteristics (i.e., limnology). Perhaps the most complete study was done between 1991 and 1996 by Gary Vineyard (University of Nevada, Reno). Much of the discussion below is taken directly from his unpublished, technical reports. Donner Lake thermally stratifies each summer at a depth of approximately 10 m. In the fall, the epilimnion deepens as the thermocline is eroded. Complete mixing typically occurs in the early winter. An ice covering can form during cold winters, but is not an annual feature. Hypolimnetic oxygen depletion is a common feature during the stratified period; however, in the autumn, hypolimnetic-dissolved oxygen concentration typically does not drop below 4 to 5 mg/L.

Vineyard (University of Nevada, Reno) classified Donner Lake on the high end of oligotrophic, approaching mesotrophic. Nutrient levels are consistent with a meso-oligotrophic status. Average concentrations for ammonium-N and nitrate-N between 1991 and 1993 were in the range of 50 and 85 μg/L, respectively, with total phosphorus at 30 to 35 μg/L. The average chlorophyll concentration from all the samples collected from 1991 to 1993 was approximately 10 μg/L; maximum chlorophyll reached levels as high as 25 μg/L. Mean Secchi depth, taken as a measure of water clarity, averaged about 8 to 9 m, with a range of 4 to 14 m. Electrical conductivity is low at 75 to 100 micro siemens (μS) with a pH near neutrality.

Meteorological Data

Data requirements. Accurate modeling of lakes and reservoirs requires reliable meteorological data near the lake surface. DLM model requires daily averages for visible and infrared radiation, air temperature, vapor pressure (which is calculated from air temperature and relative humidity), wind speed, and rainfall. It is particularly important to have good weather information for the model calibration process, to ensure that the model bases its calculations as closely as possible on the actual conditions encountered.

Installation of equipment. To obtain accurate weather information during the 2000 boating season, a meteorological station was installed at the Donner Lake public boat ramp on June 29, 2000. Instrumentation included a Kipp and Zonen CM-3 short-wave radiometer, a Met One 034A cup and vane wind sensor, and a Campbell Scientific CS500 (Vaisala 50Y) temperature/relative humidity probe. A Campbell Scientific CR-10 data logger was used to read the instruments and store the data. A 10-watt solar panel provided power to keep the battery charged.

The instruments were mounted on a pole on the boardwalk next to the lake. Instrument height was approximately 6.5 meters above the lake surface.

The CR-10 read the sensors once a minute and recorded averages every 15 minutes. Data were downloaded from the logger approximately every 3 weeks. Logger capacity was about 28 days of data. Communication with the CR-10 was accomplished with a laptop computer running the Campbell Scientific PC208W2.3 data logger communications software package.

Data retrieval. Due to the large quantity of data, the raw CR-10 data were compiled in a spreadsheet file and then exported to a Microsoft Access® database to facilitate retrieval. The average daily values required by DLM model were calculated from the 15-minute values stored in the database. Because long-wave radiation data were not collected directly, values were computed from available weather data, using a bulk equation provided by Chapra (1997).

The modeling period began on June 15, 2000, and because on-site meteorological data were not available until June 29, weather data from the UC Davis met station at Tahoe City (approximately 22 km southeast) were used as an approximation during that period. Rainfall data for the season were obtained from the Tahoe facility because no rainfall data were collected at Donner Lake.

Data analysis. The 15-minute averages recorded by the CR-10 were consolidated into a single average value for each day, for input to DLM-2002. The daily averages for July and August were also averaged to obtain average met values for the main boating season. These values are tabulated in Table 3.4, along with the 1997 values for comparison.

Table 3.4

Average meteorological values for July and August

Year	Short-wave (visible) Radiation (kJ/sq. m./d)	Long-wave (infrared) Radiation (kJ/sq. m./d)	Temp (° C)	Vapor Pressure (mb)	Wind Speed (m/s)
2000	24413	26244	16.5	7.5	1.83
1997	25554	26732	16.5	9.6	2.96

For radiation and air temperature, the two years are quite similar (< 5 percent difference). However, average wind speed in 2000 was nearly 1 meter per second (m/sec) (38 percent) lower than reported in 1997. The difference is significant, as a sensitivity analysis shows wind to have a far greater effect than any other parameter on the rate of volatilization of VOCs from the lake. The 1997 wind data were collected from the Northern Sierra Nevada Air Quality District, located in Truckee, less than 1 km from the east end of the lake.

Part of the difference may be accounted for by the off-lake locations at which the 1997 values were measured. Temperature calibration of the model against measured temperature data demonstrated that the measured wind values were too low, likely because of the shore location of the sensor. The measured values had to be multiplied by 1.2 for the model to reproduce actual temperature conditions in the lake. Thus, the measured average wind of 1.83 m/sec at the shore station was calculated to represent 2.2 m/sec on the lake, or 26 percent lower than in 1997.

Average vapor pressure in 2000 was 22 percent lower than for the same period in 1997. Because vapor pressure is a function of air temperature and relative humidity, and because average temperatures were similar for both years, the difference can be attributed to the 21 percent lower relative humidity in 2000.

A graph of daily average air temperature is presented in Figure 3.8. The salient feature of this plot is unseasonably low temperatures on both the Fourth of July and Labor Day weekends. The warmest period occurred around the end of July. Other than those excursions, temperatures were relatively constant for much of the summer season. Table 3.5 presents a compilation of the ten warmest and coldest days between June 30 and September 10.

Table 3.5

Donner Lake temperature extremes during study period

Warmest Days		Coldest Days	
Date	Temperature (°C)	Date	Temperature (°C)
08/01/2000	21.8	09/02/2000	4.9
07/31/2000	21.2	09/05/2000	5.2
08/02/2000	20.8	09/04/2000	7.6
07/30/2000	19.9	09/01/2000	7.8

(continued)

Table 3.5 (continued)

Warmest Days		Coldest Days	
Date	Temperature (°C)	Date	Temperature (°C)
08/03/2000	19.8	09/03/2000	8.3
08/04/2000	18.7	09/06/2000	8.5
07/14/2000	18.7	07/04/2000	10.7
07/29/2000	18.6	07/06/2000	11.0
08/05/2000	18.6	07/05/2000	11.0
08/06/2000	18.5	09/07/2000	11.4

Except for July 14, all of the warmest days are clustered together in late July and early August. The coolest days are grouped similarly: the first six and the last six are contiguous around Labor Day, and the rest are near July 4.

A plot of daily average wind values is presented in Figure 3.9. From June 30 to September 10, daily values were consistently near or below 2.0 m/sec. Around the Labor Day holiday, however, the first autumn storm event took place, with higher winds and sharply lower temperatures.

A tabulation of the windiest and calmest days during the boating season is presented in Table 3.6. While windy days were more scattered through the summer, they were also well represented in the two holiday weekends.

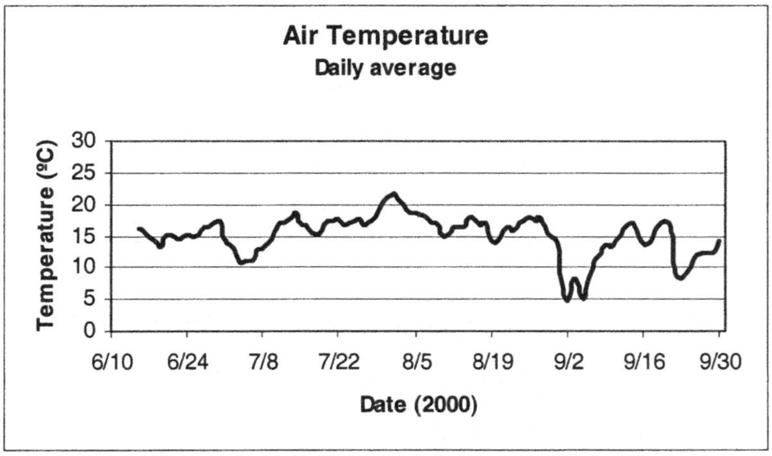

Figure 3.8 Daily average air temperature, Donner 2000

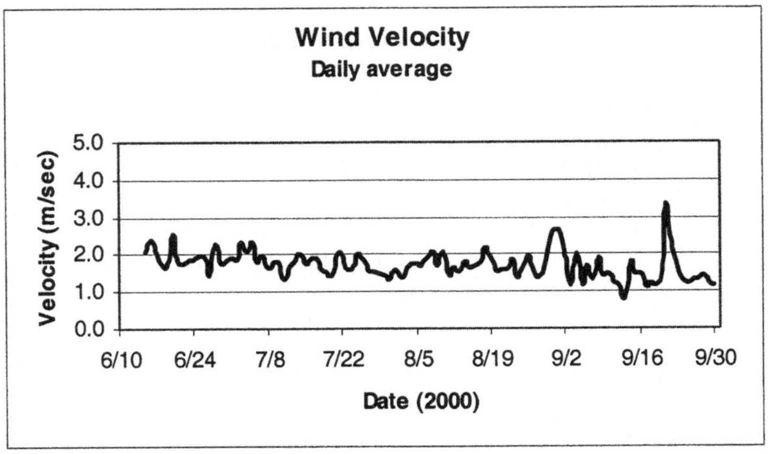

Figure 3.9 Daily average wind velocity, Donner 2000

Table 3.6

Wind extremes

Windiest Days		Calmest Days	
Date	Wind Speed (m/sec)	Date	Wind Speed (m/sec)
08/31/2000	2.7	07/20/2000	1.4
09/01/2000	2.6	07/30/2000	1.4
07/03/2000	2.3	08/24/2000	1.4
07/05/2000	2.3	08/28/2000	1.4
08/18/2000	2.2	08/02/2000	1.4
08/30/2000	2.2	07/11/2000	1.4
07/04/2000	2.1	09/07/2000	1.3
08/10/2000	2.1	07/31/2000	1.3
08/08/2000	2.1	09/05/2000	1.2
09/02/2000	2.1	09/03/2000	1.2

Input to model. The meteorological data input file for DLM-2002 was, with one exception, created directly from the measured data. The exception is vapor pressure, which was calculated from the measured air temperature and relative humidity. In addition, measured values for wind speed were passed directly to the model, but were then modified by a multiplication factor within the model code. As noted above, the multiplication factor was determined to be 1.2, and this was employed for all runs that used the 2000 Donner Lake wind data.

Year 2000 Boating Survey at Donner Lake (Census)

Purpose of Census

The purpose of the census at Donner Lake during the 2000 boating season (i.e., June 1 through September 15) was to:

- Assess use patterns for motorized watercraft.
- Assess fuel use during boating season by engine type.
- Assess amount of fuel used per activity and engine type.

Census Design

Methodology. The methodology for the census adopted included collection of data on the public boat ramp, concessionaires, and whole lake activity surveys (see Figure 3.8, above).

During the 2000 boating season, CCWD employed an intern at the Donner Lake public boat ramp to collect information on boating activity on the lake, and on types of boat engines and fuel use patterns. Detailed information was collected at the ramp by observing boats and interviewing owners during launch and recovery operations. Whole-lake-activity surveys conducted at three locations on the lake provided estimates of the amount and type of boating activity occurring concurrently (refer to Figure 3.7).

The data collected during the 2000 boating season at Donner Lake form one of the best available data sets relating boat use information to MTBE-BTEX contamination in lakes and reservoirs. The data set can provide information by date, time of day, type of engine, fuel use, etc. For the present effort, the data have value both in providing direct information about patterns of boating activity and in forming the basis for calculation of MTBE loads within DLM-2002.

On-site data were collected on paper forms that the intern then entered in electronic spreadsheets. The spreadsheet data were compiled on a weekly basis, and transmitted electronically to UCD/TRG. Data collection for boating activity on Donner Lake ended on September 29, 2000, completing 17 weeks of collection. Data were obtained for approximately

25 percent of available time slots at four launch locations and two boat rental locations. The survey collected basic information on boating activity, such as the time and date of activity, boat registration number, and engine type. Interviews with boat owners collected information on the boating activity type and duration, estimated amount of fuel used, and the location of fuel purchase. Fuel use estimates were obtained for 54 percent of the boats returning to the survey location.

Boat ramp information included a number of parameters (date, time of day, boat registration number, whether the operation was a launch or recovery, the type and Hp of the engine, and the nature [trolling, skiing, or cruising] and length of the activity on the water). Owners were also asked whether their fuel contained MTBE, and where it was purchased. Table G.1 in Appendix G is a sample listing of the data collected, as it appeared in spreadsheet form. Data were also collected from the Truckee Parks and Recreation District ramp attendant, when possible.

Activity surveys were typically conducted every 2 hours. The surveys were conducted using binoculars from three different shore locations, chosen to cover the entire lake. The surveys were conducted within as short a time as possible. Each survey involved counting the number of boats, their general propulsion type, and the activity in which each was engaged. These surveys provide a key component of the data: good estimates of the number of boats on the lake at one time.

Initial data processing. It was determined early in the data reduction process that a database would be necessary to organize and access the large amount of information collected. Microsoft Access® was chosen as a suitable software tool for this application.

Raw ramp data were first reformatted to better suit database entry. Blanks were converted to default numbers or text as appropriate, and incomplete entries were eliminated. The weekly records were consolidated into one file for the season.

Boats and engines were categorized into eight types, according to the following table:

Table 3.7

Engine types

Code	Engine Type
1	PWC, two-stroke, carbureted
2	PWC, two-stroke, direct injected
3	Outboard, two-stroke, carbureted, <=25 Hp
4	Outboard, two-stroke, carbureted, >25 Hp, <=75 Hp
5	Outboard, two-stroke, carbureted, >75 Hp
6	Outboard, two-stroke, direct injected
7	Outboard, four-stroke
8	Inboard, including stern drive

Because the intern recorded boats on both launch and recovery, and because some boats launched more than once a day, there were often multiple appearances by the same boat in one day. Only one record was retained for each boat each day, and retained recovery data where available, because they contained activity and fuel information. Accordingly, once the information was organized and ready for the database, it was pre-processed by a program that eliminated duplicate boats on the same day, and preferentially retained records with fuel use data.

With regard to the activity survey data, an assumption was made that boats were not counted twice, since there was no way to estimate the degree of such overcounting. As with the ramp data, survey data were edited to remove header information, and weekly data were compiled into one file for the season.

Database use. Creation of the database(s) allowed for selective extraction and combination of data by any of the fields. Such flexibility has proven useful both for drawing conclusions about boating patterns generally, and for preparing model input files, in particular. With respect to general boating patterns, most of the tables and plots in this report were generated from either the boat ramp or activity survey databases.

Once the database was assembled, it became a simple matter to extract data by any fields or combination of field. For example, one query generates, for each timeslot, the average number of hours spent trolling, skiing, and cruising on weekends in July. Another found days where wind velocity was greater than 4 m/sec between 9:00 a.m. and 10:00 a.m. Data can be separated by engine type, date, time of day, time of week, whether fuel data are available or not, etc.

DLM-2002 requires a file of daily MTBE loading to estimate MTBE/BTEX concentrations in the lake. Generation of that file, in turn, requires calculation of average daily MTBE loadings. While these calculations are described in Chapter 4, it is useful to note that a number of intermediate tables are developed from the database. These include the average daily number of boats engaging in each activity, the percentage distribution of engine types for each activity, the average duration of each activity, and the average fuel consumption rate for each engine type.

Census Data

The results of the census are intended to assist in the development of DLM-2002. Donner Lake was not selected because it was considered to be a lake with a "typical" boating population; indeed, readers are expected to see both similarities and dissimilarities between Donner Lake and their lake/reservoir of concern. Instead, the data from Donner Lake provided the opportunity to calibrate and verify DLM-2002. Donner Lake was selected primarily because of it had been the site of extensive studies of MTBE (Reuter et al. 1998). Indeed, at the time of this current investigation, it was one of the most extensively studied lakes in the world with regard to sources, transport and fate of MTBE, including a modeling effort (McCord and Schladow 1998).

MTBE (and BTEX) loading from marine engine emission on a daily basis throughout the 2000 boating season were derived from the boat activity survey. The model, in turn, predicts in-lake concentrations of MTBE (and BTEX) based on loading, along with what is known about the fate and transport of MTBE (and BTEX) in lakes and reservoirs.

Boat Use by Numbers

Boat use was estimated for each of the 107 days during the 2000 boating season (June 1 to September 15). The total number of motorized watercraft using Donner Lake during that period was approximately 8,825. A plot of the estimated daily boat use revealed several interesting features (Figure 3.10). A clear weekday versus weekend pattern was evident, especially during the most active boating months of July and August. The weekend increase in boat use was reflected as spikes in increased boat use ranging from 150 to 250 boats per day, up from the weekday baseline of 60 to 70 boats per day. Boat use dramatically declined in September following the Labor Day weekend to less than 40 boats per day on weekends. Peaks in boat use were not common in June, except for a single Saturday-Monday period in mid-June, when daily maximum use reached approximately 150. There are no boating data from before June 17. Data for the first 2 weeks of June are estimated.

The overall seasonal pattern of MTBE concentration in Donner Lake resembled this distribution of boats (Figure 3.10).

Total weekend boat use was 33 percent higher than weekday boat use for the period of record (Table 3.7). The weekday versus weekend difference was most dramatic in July, when weekend use was almost double weekday use. In August, however, the weekday and weekend use was nearly identical. In June, just prior to the main boating season, total weekly boating activity was moderate (approximately 1,900), with little difference between the weekend and weekday periods. Total weekly use in July and August was much higher, at 3,200 to 3,400 boats per month.

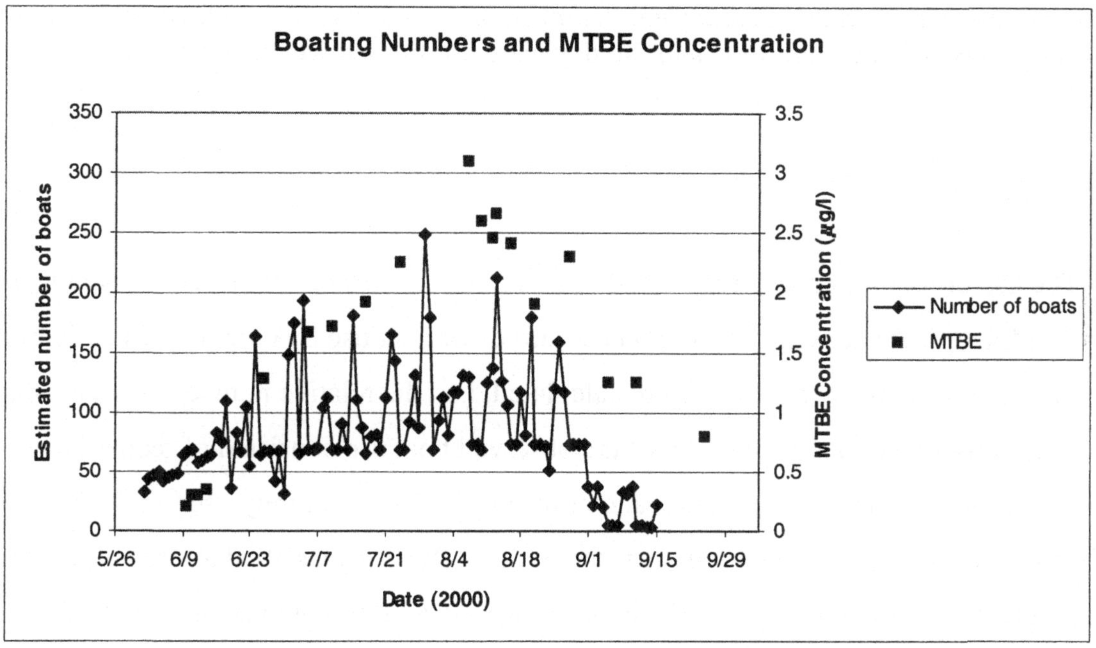

Figure 3.10 Estimated daily boat use and measured MTBE at Donner Lake

Mean daily boat use showed the elevated weekend use even more dramatically. From June to August, mean daily use on weekdays ranged from 58 to 83 boats; by September, this value declined to a mean of only five boats per day on weekdays (Table 3.8). In contrast, values on the weekend days ranged from 71 boats per day in June, to 137 and 135 boats per day, respectively, in July and August. By September, following the Labor Day weekend, daily boat use on the weekend was down to 40 boats per day. Combined mean weekday and weekend boating activity by month was 64, 108, 103, and 24 boats per day for June, July, August, and September, respectively.

Table 3.8

Monthly boat use at Donner Lake during the 2000 boating season*

Month	Weekday Use	Weekend Use	Total Use
June	995(58)	924 (71)	1,919 (64)
July	1,162 (77)	2,192 (137)	3,354 (108)
August	1,582 (83)	1,617 (135)	3,199 (103)
September	34 (5)	315 (40)	353 (24)
TOTAL	3,777 (65)	5,048 (103)	8,825 (82)

* Values represent total number of watercraft with marine engines on the lake during the period denoted. Numbers in parentheses represent the mean daily boat use for weekday and weekend days, respectively. Friday, Saturday, and Sunday were considered weekend days, as were holidays.

Boat Use by Activity

The use of motorized watercraft in Donner Lake was further delineated to include boat use profiles for specific activities. Based on our survey of boat use by numbers, and additional census' for activity, activity fractions were calculated, i.e., the relative number of boats engaged in a specific activity (see Table 3.9). All of the observed boat use consisted of recreational activities including fishing, water skiing, and cruising. PWCs were rarely used for fishing or skiing, so for the purpose of this discussion, the use of a PWC is classified as cruising. Fishing generally refers to trolling, but includes the use of an engine to travel to a particular location. Skiing refers solely to traditional water skiing, including wakeboarding and other activities that involve towing. Cruising includes all activity that is not specifically fishing or skiing. These fractions are important in calculating total fuel use and MTBE/BTEX emission, because each activity requires different engines that have different operating characteristics vis-à-vis fuel consumption.

Table 3.9

Activity fractions for 2000 boating season at Donner Lake*

Month	Weekend			Weekday		
	Troll	Ski	Cruise	Troll	Ski	Cruise
June	0.14	0.33	0.53	0.19	0.27	0.54
July	0.14	0.33	0.53	0.19	0.27	0.54
August	0.12	0.38	0.49	0.12	0.36	0.52
September	0.55	0.12	0.33	0.17	0.31	0.52

* Values represent relative number of motorized watercraft engaged in the specified activity.

The relative number of boats involved in trolling, skiing, or cruising activities generally remained consistent in July and August. Since no activity estimates were made in June, the July fractions were used as reasonable estimates. Furthermore, this consistency of activity was also found when comparing weekdays to weekends. Trolling was the least common activity from June-August. On the weekends, trolling accounted for 12 to 14 percent of boating use (Table 3.9); on the weekdays, this value increased somewhat (12 to 19 percent). This trend was due to the fact that people using low-Hp engines to troll during fishing represented a high portion of the boating population during the "slower" weekday periods. On the weekends, more boaters were engaged in skiing and cruising activity. From June-August, water skiing was the next most popular boating activity, accounting for 27 to 38 percent of the total boat use. On the weekends, the activity fraction was relatively uniform at 0.33 to 0.38, whereas on weekdays, values were just slightly lower, at 0.27 to 0.36. Cruising accounted for the largest portion of boating activity, with slightly greater than 50 percent of the motorized watercraft involved in this activity.

In September, following the height of the summer boating season, there was a dramatic shift in boating activity. On weekends, trolling was the most prevalent activity with 55 percent of all boating engaged in this activity. Skiing use was down to only 12 percent, with cruising reduced to 33 percent. This represents the post-Labor Day change away from recreational boating to fishing. During weekdays in September, the pattern of boat activity was very similar to that seen during the period June-August.

Figures 3.11 and 3.12 depict the strong "time of day" dependence in boat activity. Trolling, skiing, and cruising activity during the high boating months of July and August was evaluated on the basis of what time during the day these activities were observed. As described above, this was done separately for weekdays and weekends. In general, daily use patterns were very similar during these two time periods; however, all types of boating activity were elevated on the weekends, as discussed above. Trolling, associated with fishing, was most common during the morning hours, especially between 7:00 a.m. and 11:00 a.m. A decline in this activity was observed after the morning hours; by afternoon, trolling was negligible. Trolling remained low in the early evening hours between 5:00 p.m. and 9:00 p.m.

Skiing showed a distinct diurnal pattern, with maximum activity between 9:00 a.m. and 3:00 p.m., peaking in late morning-early afternoon. While there was a minimum in skiing from 7:00 p.m. to 9:00 p.m., coinciding with the onset of dusk and eventual darkness, skiing nonetheless was observed throughout the daylight hours. Cruising dominated boating activity primarily in the afternoons, with most activity occurring between 11:00 a.m. and 5:00 p.m.

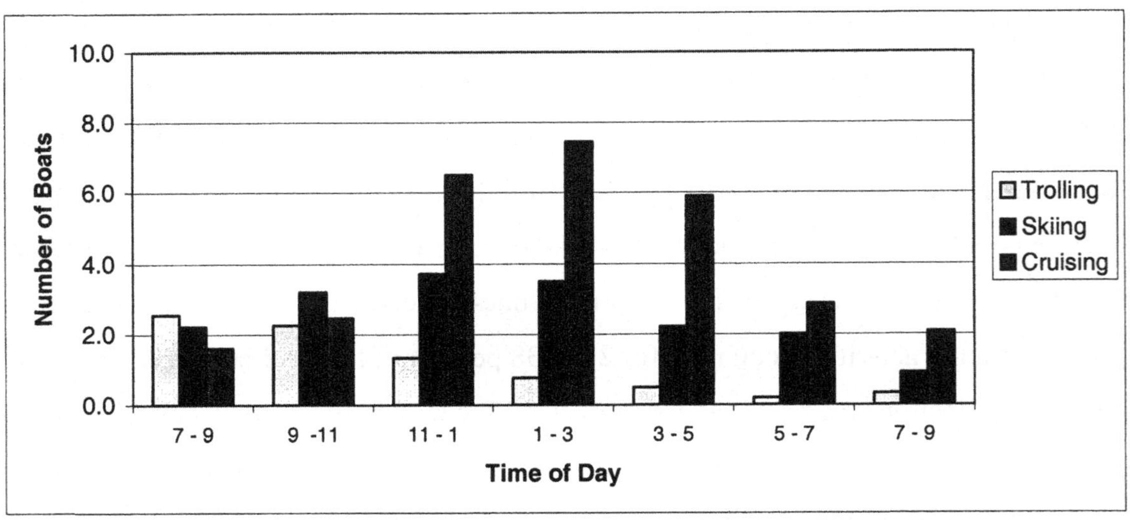

Figure 3.11 Average weekend boating activity July and August 2000 (Refer to Table 3.10, below, for engine code descriptions)

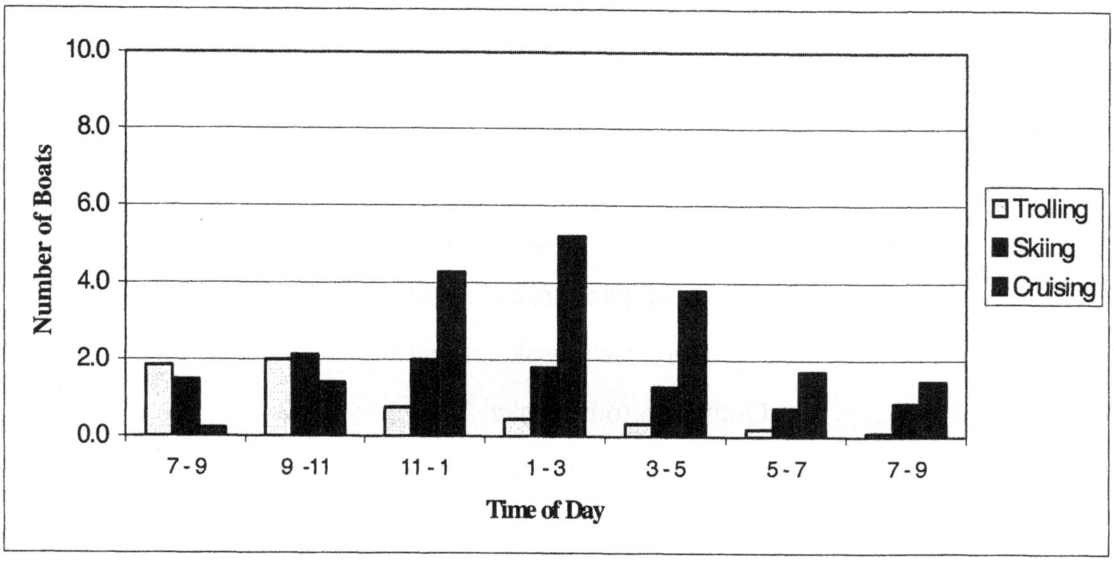

Figure 3.12 Average weekday boating activity July and August 2000 (Refer to Table 3.10, below, for engine code descriptions)

The survey was also used to estimate the duration of each of the primary boating activities. These data were used to calculate hours of operation for the various engine types (see discussion below). The calculated average duration for trolling, based on boat operators' reports, was 3.8 hours/day/boat. Trolling had the highest duration followed by cruising at 2.6 hours/day/boat and skiing at 2.4 hours/day/boat. Use of PWCs averaged 3.2 hours/day/boat.

Engine Type

Based on the Donner Lake boating census, eight types of marine engines were identified as being used in that lake/reservoir:

Table 3.10

List of marine engines and codes

Engine Code	Type
1	PWC, two-stroke, carbureted
2	PWC, two-stroke, direct injected

(continued)

Table 3.10 (continued)

Engine Code	Type
3	Outboard, two-stroke, carbureted, □25 Hp
4	Outboard, two-stroke, carbureted, >25 Hp, <75 Hp
5	Outboard, two-stroke, carbureted, □75 Hp
6	Outboard, two-stroke, direct injected
7	Outboard, four-stroke
8	Inboard, four-stroke, including stern drive

Accounting for the estimated boat usage on Donner Lake from June to September 2000, the dominant marine engine type was the four-stroke inboard, including stern drive. These types of watercraft are primarily used for water skiing and cruising. Of the estimated 8,825 boats to have used Donner Lake, an overwhelming 60 percent were of this engine type (see Figure 3.13). Combined, the four-stroke engines comprised approximately 65 percent of the watercraft with some variation of the two-stroke technology accounting for the remaining 35 percent. PWCs, expressed as the sum of engine type 1 and 2 categories, were represented by an estimated 728 watercraft, or 8 percent of the boating population. After the four-stroke inboards, the smaller two-stroke carbureted outboards less than 25 Hp were the next most common engine, comprising 12 percent of the total boating population. When the data were analyzed by month, the relative percentage of engine types from June to August reflected the annual, cumulative data presented here.

Fuel Consumption

Fuel consumption was estimated on the basis of month, weekend versus weekday, activity type, and engine type. Calculated values for hours of operation were multiplied by specific fuel consumption rate to obtain gallons of fuel used. Fuel consumption rates were obtained during the summer survey in Donner Lake as boaters returned to the ramp to exit.

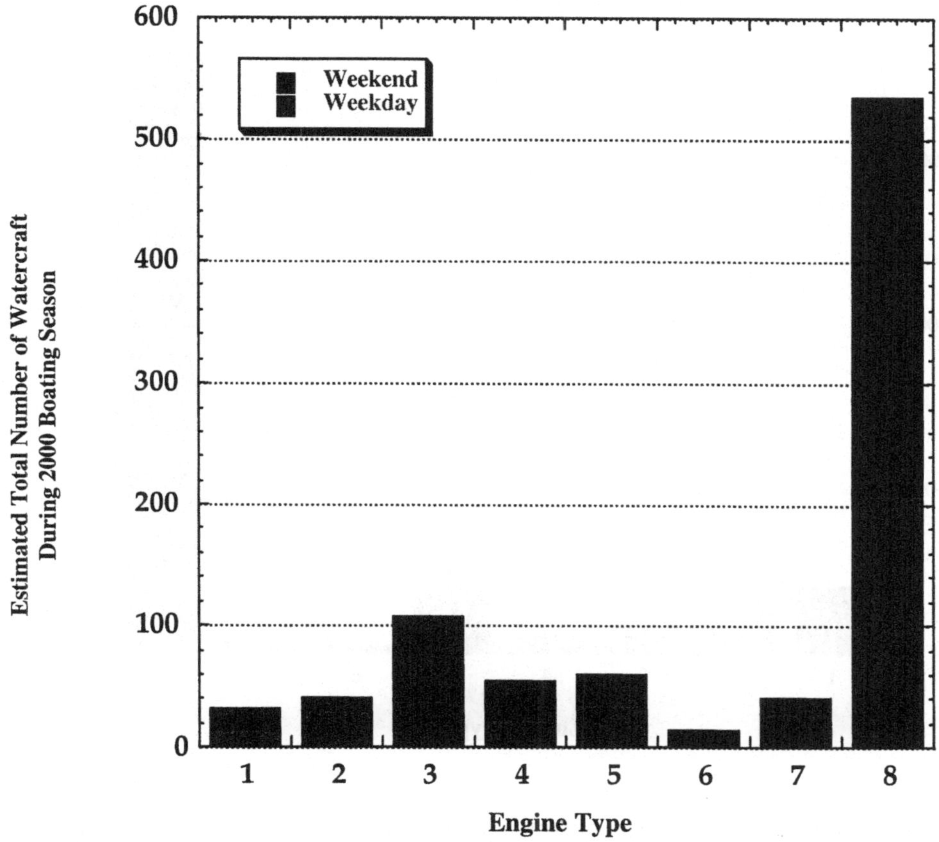

Figure 3.13 Estimated boat usage on Donner Lake by engine type during the 2000 boating season (refer to Table 3.10 for engine code descriptions)

A total of 26,564 gallons of fuel were used by all motorized watercraft at Donner Lake during the 2000 boating season, June to September. This resulted in an overall mean consumption of 3 gallons per boat per outing. Of this fuel use, the four-stroke engine boats were the dominant consumers, accounting for 70 percent, especially the inboard engines (i.e., engine type 8) (see Figure 3.14). Two-stroke technologies used 30 percent of the fuel with PWCs accounting for 15 percent. Outboard, carbureted two-stroke, >75 Hp engines consumed 10 percent of the total fuel.

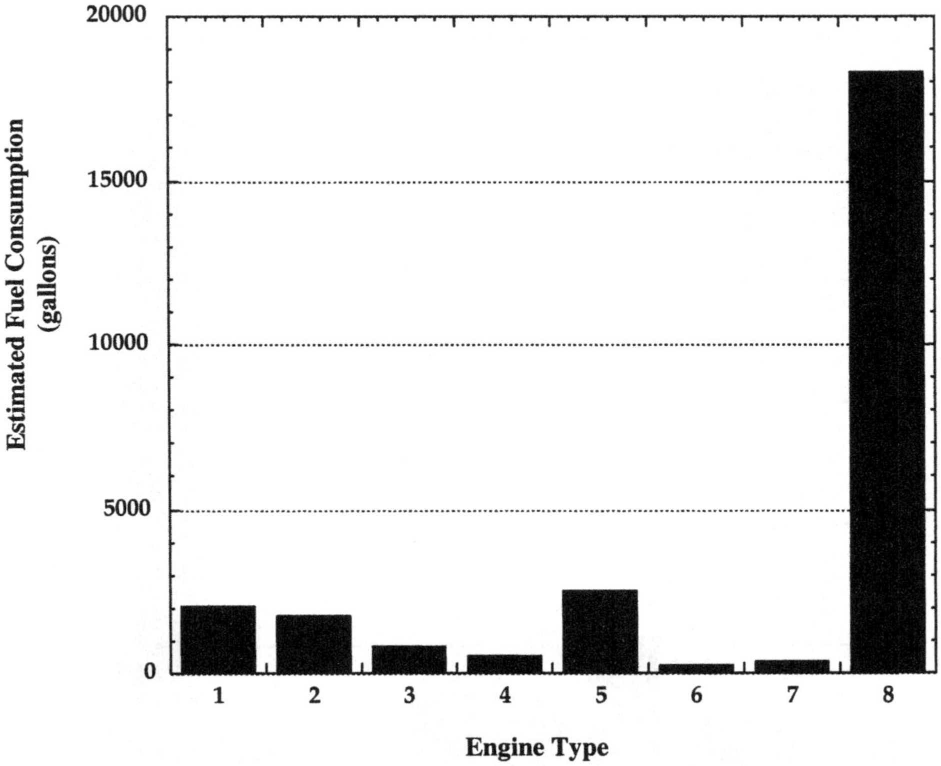

Figure 3.14 Estimated engine fuel consumption by engine type for all motorized watercraft at Donner Lake during the 2000 boating season. (Refer to Table 3.10 for engine code descriptions)

Release of MTBE, Benzene, and Toluene to Lake Water

Based on the calculated gallons of fuel consumed by motorized watercraft, and test data on emission rates of VOCs from the various types of marine engines (Table 3.2), an estimate was made of the volume of MTBE, benzene, and toluene released to Donner Lake (Table 3.11). Values are presented by engine type, and represent a cumulative load for the entire boating season.

Total MTBE load to Donner Lake during the 2000 boating season amounted to 123 gallons. Two-stroke engines accounted for 118 gallons, or 95 percent, of the total load, even though these engines only consumed 30 percent of the fuel. The four-stroke technologies, while consuming 70 percent of the fuel were responsible for only 5 percent of the MTBE released to the lake. Similar findings were reported for Lake Tahoe (Allen et al. 1998). Within the two-stroke category, PWCs released 33 gallons, or 27 percent, of the total; the carbureted, low Hp

(□25 Hp) engines contributed 25 percent of the total load; and the carbureted engines >75 Hp contributed 32 percent of the total. Despite the fact that the carbureted, two-stroke engines consumed less fuel, their contribution to release to the lake was more significant because of the higher release rates associated with these engines (i.e., 10 to 31 percent of fuel passes though engine unburned). The same pattern for two-stroke versus four-stroke emission was seen for both benzene and toluene, which had loadings of 15 gallons and 59 gallons, respectively, for the boating season.

Table 3.11

Calculated volume (gallons) of VOCs entering Donner Lake from marine engine emissions*

Engine Type	1	2	3	4	5	6	7	8	Total
MTBE	26	7	31	14	39	0	0	5	123
Benzene	3	1	4	2	5	0	0	1	15
Toluene	12	3	15	7	19	0	0	2	59

* Based on engine type and cumulative for entire boating season. MTBE is taken to be 14 percent of gasoline by volume, benzene is 1.7 percent by volume and toluene 6.7 percent by volume.

MTBE/BTEX Water Quality Monitoring Program

This section presents a brief description of the purpose of the monitoring program to set the context for the guidance manual, together with the salient results and conclusions. For more details on the sampling design and results of the program, please refer to Appendix H.

Purpose of Sampling Program

The primary purpose of the monitoring program was to obtain field data to allow for verification of DLM-2002 during the 2000 boating season at Donner Lake. Prior to using the model for guidance purposes, it was necessary to determine if the calibration was accurate (i.e., if the results mimicked actual ambient conditions). To do this, an extensive field-monitoring program for MTBE and BTEX was carried out to provide data on a number of scales, including seasonal, weekly, daily, synoptic, and distribution, with depth from surface to bottom. Since boating occurs principally between Memorial Day (May 29, 2000) and Labor Day (September 4, 2000), sampling was clustered within this timeframe.

A further benefit from the monitoring program was that it provided a good comparison with 1997, when a similar comprehensive study of MTBE was done in Donner Lake (Reuter et al. 1998). With this additional year of data, conclusions can be drawn regarding impacts of management differences between these two years (use of MTBE-free fuel by boat concessionaires in 2000), as well as the possible influence of the difference between boating and weather conditions between these two years. As noted elsewhere in this report, the original DLM-MTBE model was calibrated and applied to Donner Lake using the 1997 data.

Sampling Design

Site selection was facilitated by results from the 1997 Donner Lake monitoring. The 1997 results showed that MTBE was distributed evenly throughout much of the surface area of Donner Lake. On those dates when MTBE exceeded a concentration of 1 μg/L, the coefficient of variation (standard deviation ÷ mean) for measurements taken at three sampling locations from west to east was low and typically <10 percent. The lake is characterized by an oval-shaped shoreline with no sheltered embayments or enclosed marinas. The long, east-west axis also aligns with the prevailing winds, contributing to the spatially well-mixed conditions.

Based on these findings, two primary sites were chosen for sampling (refer to Figure 3.7). These sites were also sampled in 1997. Based on the results of the 1997 study, the sampling depths for the present study in 2000 were fixed to focus on the epilimnion layer.

The Donner Lake 2000 boating season sampling schedule was designed to capture the three primary temporal scales, which the model was targeted to predict. These include (1) seasonal time scale; i.e., looking at changes in MTBE/BTEX as the boating season progressed during its typical Memorial Day to Labor Day period, (2) weekly time scale, i.e., what happens to MTBE and BTEX during weekend boating activity and the intervening weekday period between, and (3) daily time scale, i.e., did concentrations vary with time of boating activity on a given day.

While the 1998 version of DLM was able to predict on each of these time scales (McCord and Schladow 1998), this "first-generation" model could only be validated for the annual-seasonal scale, because data for the other times scales were unavailable. As part of the current study, an expanded monitoring program was completed in which 156 samples were taken in 2000 and analyzed for MTBE and the BTEX compounds (benzene, toluene, ethylbenzene, and xylene). This allowed us to more adequately validate the updated version of the model. Note that the sum of the data points available for analysis for the four categories presented below is 169. The reason for the discrepancy between this value and the actual number of samples collected lies in the fact that some of the collected samples could be analyzed as part of more than one of the categories.

Methodology

Sampling for MTBE in Donner Lake was conducted from either a non-motorized kayak or a 14-foot aluminum fishing boat powered by a 15 Hp, four-stroke outboard motor. To ensure no contamination from the motorized vessel, all sites were approached from down wind, and the boat was anchored on location.

Water samples were collected from discrete depths using a Kemmerer well sampler (Table 3.12). In situ temperature profiles were obtained at each station using an Richard Brancker Research, Ltd. (RBR) temperature and depth data logger. Surface reference temperatures were taken with a hand-held thermometer at each sampling location.

The analysis was carried out at CH2M Hill's Corvallis Applied Sciences Laboratory (CVO ASL). Gas chromatography techniques used for the analysis and quality control procedures are based on USEPA Method 8000B, revision 2.0. Purge-and-trap concentration procedures are based on USEPA Method 5030B, revision 2.0. Analysis procedures are based on USEPA Method 8260B, revision 2.0. All of these methods are published in USEPA SW846, Test Methods for Evaluation of Solid Waste, May 1997. Field and Laboratory Quality Assurance/Quality Control (QA/QC) procedures were adopted to ensure accuracy and precision in data generated from the program. All of the aforementioned methods were chosen because they are USEPA standards. More information on test methods and laboratory quality control is included in Appendix H.

Results

Frequency of Detection. Of the 156 total number of samples analyzed for MTBE and BTEX, 89 (57 percent) contained MTBE at a level greater than, or equal to, the 0.2 μg/L detection limit. However, of the 88 samples taken in the epilimnion, 86 (98 percent), or nearly all of the samples, contained MTBE at a level greater than, or equal to, the 0.2 μg/L detection limit. At no time did MTBE exceed the California secondary-drinking water standard of 5 μg/L. The BTEX compounds were found much less frequently. Benzene rose above its 0.2 μg/L detection limit only on a single occasion, when a concentration of 0.3 μg/L was measured on-site during the mid-August synoptic survey. This accompanied the highest single MTBE value found in this study, 3.6 μg/L. For comparison, the California primary-drinking water standard for benzene is 1 μg/L.

Toluene was the most commonly measured BTEX compound found in 23 of the 88 eplimnetic samples (26 percent). Toluene typically ranged from 0.2 to 0.6 μg/L; however, a single higher value of 1.5 μg/L was found, again on-site during the mid-August synoptic. Even the largest toluene concentration observed was much lower than the California primary-drinking water standard of 150 μg/L and the USEPA taste and odor threshold of 42 μg/L. Similar to benzene, ethylbenzene was only found on a single occasion, but at the 0.2 μg/L detection limit. For comparison, the California primary-drinking water standard is 700 μg/L, while the USEPA taste and odor threshold is 29 μg/L.

Table 3.12

Reporting levels or analytical limits of detection at Donner Lake

MTBE	0.2 μg/L
Benzene	0.2 μg/L
Toluene	0.2 μg/L
Ethylbenzene	0.2 μg/L
m,p – xylene	0.4 μg/L
o – xylene	0.2 μg/L

Xylenes were found in only three of the 88 epilimnetic samples (3 to 4 percent). For two of these samples, only o-xylene was seen (both at the 0.2 μg/L detection). The synoptic (site f) sample contained both m,p-xylene (0.9 μg/L) and o-xylene (0.4 μg/L). Again, these few observed concentrations were well below the California primary-drinking water standard of 1,750 μg/L and the USEPA taste and odor threshold of 17 μg/L.

Distribution with Depth: The transport of MTBE to depth was significantly retarded by the thermocline. This was identical to the observation made in 1997 (Reuter et al. 1998).

Representative mid-monthly profiles are given in Figure 3.15. The confinement of MTBE to epilimnetic waters is obvious, with no MTBE found at a depth 2m below the thermocline.

The data clearly show that the thermocline acted as an effective barrier to the downward movement of MTBE. None of the BTEX compounds was found below the thermocline. The 14-m depth represents a position in the water column 2-m below the thermocline. On September 11 the actual sampling depth was 16 m; on October 23 it was 21 m, reflecting the seasonal deepening of the thermocline.

Seasonal. Twenty-three individual sampling dates were available to construct seasonal changes in MTBE concentration (Figure 3.16). These included the routine seasonal sampling plus the two intensive weekly sampling dates. Data are plotted as mean epilimnetic levels.

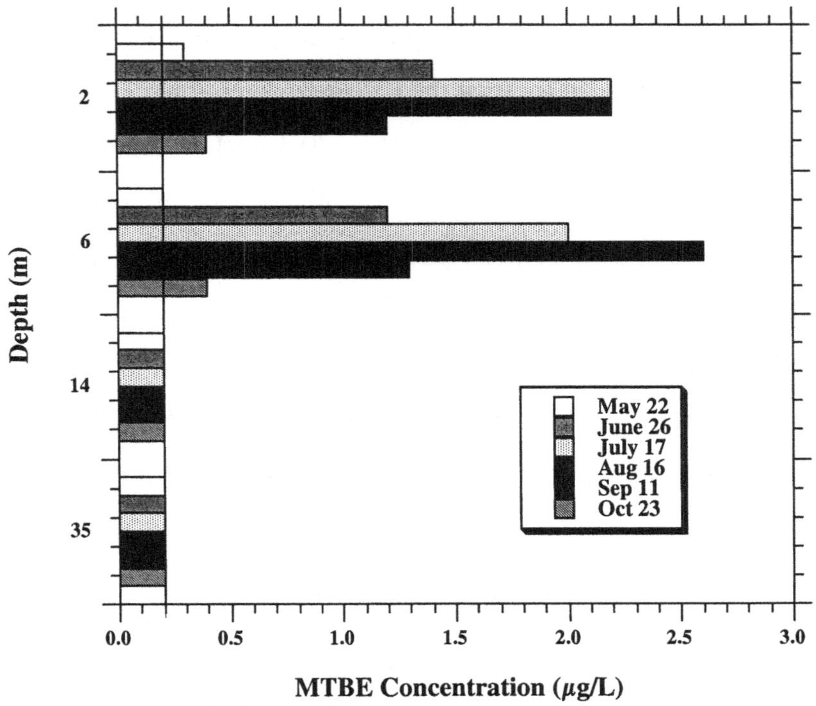

Figure 3.15 Depth distribution of MTBE in Donner Lake

The distribution of MTBE in Donner Lake followed a clear cyclical pattern with a number of distinct periods. This pattern has been previously seen at a number of lakes and reservoirs, including Lake Tahoe (Allen et al. 1998), Lake Perris in Southern California (Dale et al. 2000), and the previous Donner Lake study (Reuter et al. 1998). During the 2000 Donner Lake study, MTBE was low, albeit above detection, from late May to mid-June (0.2 to 0.4 μg/L). Beginning in late June, concentration began to steadily increase until August 7 when the maximum level of 3.1 μg/L was measured. This corresponded to the onset of boating. A significant difference between 1997 and 2000 was that in 1997, an 8 to 10 μg/L increase in MTBE was seen between July 1 to 7, associated with heavy boating over the Independence Day holiday. In 2000, no such spike in MTBE was seen and the absolute concentrations of MTBE were much lower (e.g., maximum MTBE in 1997 was 12.1 μg/L, but only 3.1 μg/L in 2000 [see Conclusions in this report]).

Weekly. Sampling for weekly changes in MTBE was done on two occasions: mid-June, when boating activity is low, and mid-August, when activity is seasonally high. The BTEX compounds were not above their detection limits for either of these periods; consequently, the presentation of data is limited to MTBE. Similarly, concentrations in mid-June were very low.

Despite the fact that all the epilimnetic samples were above detection, the range and the absolute values were only 0.2 to 0.4 μg/L – too small to draw conclusions.

The mid-August data showed daily changes in MTBE concentration; albeit, these were smaller than the more general seasonal changes. By adding a sampling profile done on the previous Monday to the Thursday to Wednesday sequence, the program was able to detect differences during the week (Figure 3.17). The observed drop in MTBE was most likely the result of minimum boating activity (new inputs), combined with loss due to volatility.

Although some of the day-to-day changes detected during the week of August 7 to 16 were small, clear differences were evident and qualitatively explainable by boating activity. The largest change in MTBE was the 0.75 μg/L, or 25 percent decrease from Monday to Monday during this week. As shown by the data defining seasonal changes (Figure 3.16), the variations in MTBE over the period August 7 to 16 are clearly discernible on this plot.

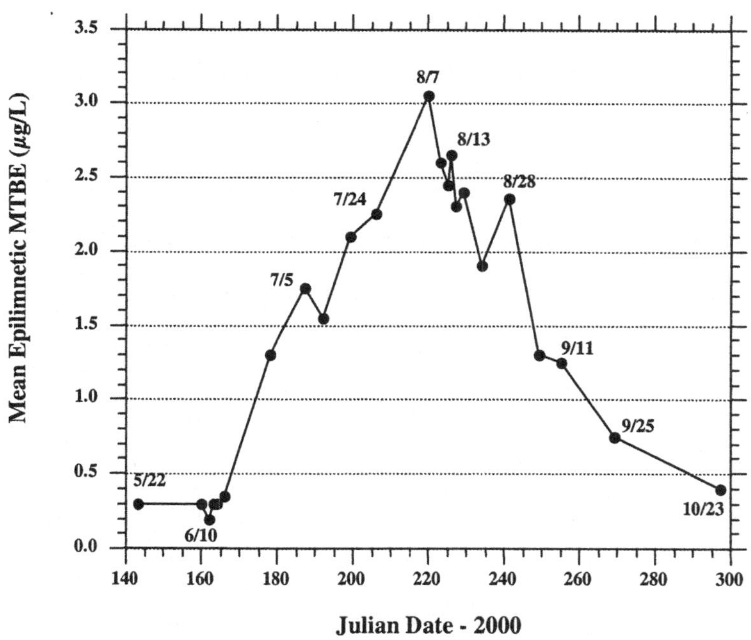

Figure 3.16 Seasonal distribution of epilimnetic MTBE in Donner Lake, May 22 – October 23, 2000

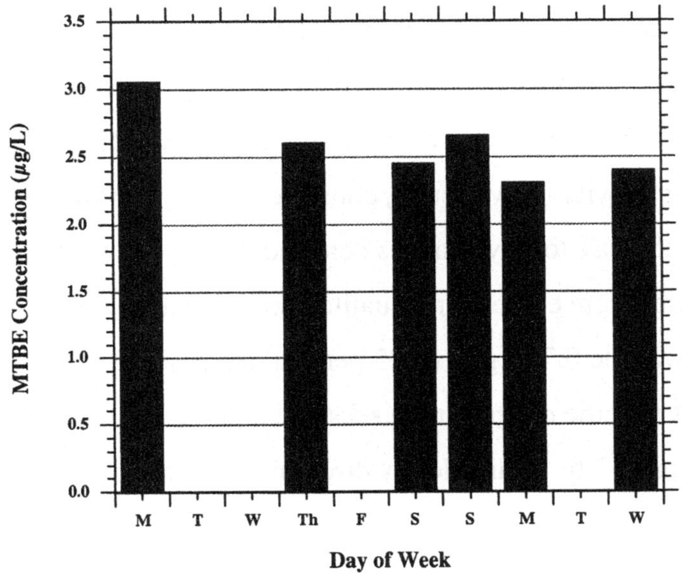

Figure 3.17 Change in MTBE concentration at Station 4 over a 10-day period in mid-August

In the only published study that the authors of the program could find which investigated MTBE variation within a week, Baehr and Reilly (2001) reported on a 1999 study in Lake Hopatcong, New Jersey. They sampled five times between Aug 26 and September 3 and four times between September 10 and September 20. MTBE was typically higher on Sundays and Mondays. The authors presumed this was due to an increase in weekend boat traffic. Absolute concentrations varied from 5.7 to 18.4 μg/L (at 3-foot depth) with increases over the weekends ranging from 1.2 to 2.7 times the initial value. Declines from Monday to Wednesday were 17 percent and 54 percent.

Daily. Monitoring for daily changes in MTBE was done on a mid-August weekend when boating activity was expected to be seasonally high. As presented above, sampling was done four times between 7:00 a.m. and 6:55 p.m. to capture the daily pattern of boat use in Donner Lake. The analytical results suggest minor differences in MTBE concentrations on this time scale (at least on the day of sampling).

Synoptic. As part of the effort to determine spatial variation between locations, a statistical comparison of the epilimnetic depths at Station 4 versus Station 5 is presented. There were 17 occasions when at least one of the MTBE concentrations was greater than the 0.2 μg/L limit of detection.

The synoptic survey of Donner Lake was performed in mid-August, on a weekend when boating activity was expected to be elevated. The sampling began at 12:20 p.m. and ended approximately two hours later at 2:25 p.m. Both MTBE and toluene were above detection, thereby allowing for an analysis of each.

MTBE concentrations around the lake varied by a maximum of 1.2 μg/L during the synoptic survey. Both the synoptic sampling and the Station 4 versus Station 5 comparison indicated little spatial variation in MTBE and BTEX in Donner Lake, thereby supporting previous conclusions (Reuter et al. 1998).

Summary of Donner Lake Monitoring Data

(1) BTEX was found only infrequently in the Donner Lake, year 2000 monitoring. Values rarely exceeded two to three times their analytical limits of detection. At no time did BTEX approach a concern, vis-à-vis drinking water standards. Given the higher volatility of BTEX relative to MTBE, this was not surprising. The lack of BTEX in Donner Lake does not imply that it this is not an important fuel constituent. According to Fiore et al. (1998), toluene accounts for 8 percent of California Certified Fuel while total xylenes account for 11 percent (MTBE accounts for 14 percent). For a two-stroke, 80-Hp carbureted PWC in Lake Tahoe after 30 minutes of operation, toluene and total xylenes exhausted into the water were measured at 17 μg/L and 23 μg/L.

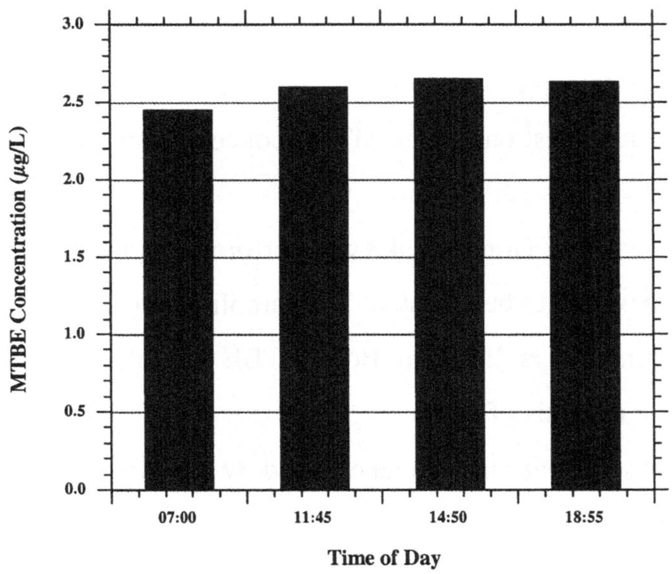

Figure 3.18 Daily changes in MTBE in Donner Lake as measured on a mid-August weekend

Not only are these concentrations relatively high, their ratios to MTBE were similar to that in the raw fuel. These same authors reported toluene and xylene levels in an area of high PWC use in the range of 5 to 10 μg/L. Similarly Dale et al. (2000) found maximum toluene levels of 6 to 7 μg/L in Lake Havasu, with xylene at about 3 μg/L. Because BTEX can be volatile, the elapsed time since discharge/loading will affect its concentration of surface lakes and reservoirs. Since sampling cannot always be that responsive to boating activity or other sources, the use of a model may be the best way to predict expected BTEX levels. Even the frequent, below-detection levels found at Donner Lake also verified the model.

(2) In Donner Lake, the spatial distribution of MTBE was relatively uniform. This was especially true in the open waters. Along the shoreline, concentrations were slightly higher in an area of extensive PWC use, but not more than 1 μg/L above the other nearshore sites. This substantiates our use of the 1-dimensional DLM-MTBE model. Lakes and reservoirs with irregular shorelines (e.g., bays, coves, arms) might be expected to demonstrate more spatial variation depending on size, volume, mixing patterns and currents.

(3) As shown for a number of lakes and reservoirs, the summer thermocline in Donner Lake acted as an effective barrier to the downward transport of MTBE and BTEX to deeper depths. MTBE was rarely found below the thermocline, suggesting that modifying the depth of withdrawal for drinking water could be an effective management option, provided other conditions for water quality and treatment were met (e.g., low suspended solids, reduced taste and odor compounds). When the lake mixes in the fall / winter, MTBE will temporarily be transported to deeper depths.

(4) The seasonal distribution of MTBE in Donner Lake followed a distinct cyclical pattern with a steady increase in mean epilimnetic concentration between the end of June and early August. Peak MTBE concentration occurred on August 8, at 3.1 μg/L. Elevated levels were not associated with either the Independence Day or Labor Day holidays. Cold weather conditions at both these times resulted in reduced boating. The calculated half-life for MTBE at the end of the effective boating season was 22 days.

(5) Differences in MTBE concentrations were detected over the course of a 10-day, mid-summer period. Levels during, or immediately following, the weekend were higher, with 10 to 20 percent changes found over this period. Changes measured over this time scale were evident on the plot of seasonal patterns. Daily changes in MTBE were not pronounced. The largest difference was found when MTBE rose from 2.45 μg/L at 7:00 a.m. to 2.65 μg/L at 2:50 p.m.

CHAPTER 4
MODEL CALIBRATION AND VERIFICATION

INTRODUCTION

The object of the model calibration and verification process was to gain confidence that the model can accurately represent MTBE and BTEX concentrations in reservoirs. Ideally, accurate field measurements of model input variables and VOC concentrations are collected for a particular reservoir. Calibration involves making adjustments, if required, to calibration variables and input data such that the output VOC levels are close to those measured in the field. Verification is the process of applying the calibrated model to different situations for which inputs and outputs are known, and seeing that the model can reproduce measured VOC levels. At that point, the model can be applied to a variety of situations, with confidence that the model outputs will approximate real VOC concentrations for a particular set of input conditions.

PRELIMINARY CONSIDERATIONS

Quality of Input Data

As with any model, the reliability of the DLM-MTBE output is directly related to the quality of the input data. A substantial effort was made to collect a good set of input data at Donner Lake in 2000, particularly with respect to boating patterns. The result was a well-quantified data set, with on-site meteorological measurements and detailed boating use information.

Boating Data

Collection of boating activity data for Donner Lake, California began on June 17, and ended on September 29, 2000, with the bulk obtained during the peak boating months of July and August. As discussed in Chapter 3, boating activity survey data were obtained for approximately 25 percent of the available time slots during the months of July and August. Information obtained at the Donner Lake public boat ramp included observable quantities such as time, date, boat registration number, and engine type. Boat owners were interviewed to obtain information about the type and duration of activity, the amount of fuel used, and the location of fuel purchase. Fuel use estimates were obtained for approximately 50 percent of the boats returning to the survey location.

Some types of boating data were more reliable than others. Engine type information was available by inspection at the ramp and is considered accurate. Fuel use and activity length were based on boat owner estimates and are assumed to be reasonable. Some fuel use estimates appeared to be low, given the engine and activity type. However, attempts to remove questionable data proved problematic, and in the end the full set of values was used.

Meteorological Data

Meteorological data were collected at a station located on the shore of Donner Lake. The station was installed on June 29, 2000 and removed on November 16, 2000. Data collection was continuous over the period. Values were measured every minute, and averages were recorded at 15-minute intervals. No problems were encountered with the equipment, and the data set is considered to be of high quality.

Inflow and Outflow

Donner Lake outflows consist of Donner Creek, and an outlet pipe used for local water supply. A single daily value of 2,500 cubic meters per day (m3/day) was assumed for the supply outlet, the value used in 1997. Daily values for lake surface elevation and Donner Creek stream flow were obtained from the USGS web site. The Donner Creek gage is USGS site number 10338500. Daily inflow data were not directly available, and was inferred using the elevation and outflow data.

MTBE and Temperature Profile Information

An extensive MTBE/BTEX water-quality monitoring program was carried out at Donner Lake during the May through October 2000 boating season. Its purpose was to obtain accurate MTBE/BTEX concentration data to calibrate and verify DLM-2002. Appropriate protocols were followed both in field sampling and laboratory analysis. Laboratory detection limits were 0.2 μg/l for MTBE, benzene, toluene, ethylbenzene, and o-xylene; and 0.4 mg/l for m- and p-xylene. Details of the sampling program are described in Chapter 3 and Appendix H. The water-quality sampling results can be considered to accurately represent conditions in Donner Lake.

Vertical water temperature profiles were collected at the time of sampling, in order to provide calibration data for the hydrodynamic portion of DLM-MTBE. The RBR temperature and depth data logger collected rapid measurements as it was lowered through the water column, providing clear delineation of the epilimnion, thermocline, and hypolimnion.

Engine Emission Rates

Engine emission rates used for modeling are summarized in Table 4.1, and represent the best data available as of February, 2001. A more complete discussion of engine emissions is presented in Chapter 3, and summarized in Table 3.3. The emission rates were developed based on MTBE, and were used for modeling all five VOC compounds.

Table 4.1

Engine emission rates

Engine Code	Engine Type	Emission Rate, %
1	PWC, two-stroke, carbureted	9
2	PWC, two-stroke, direct injected	3
3	Outboard, two-stroke, carbureted, □25 Hp	28
4	Outboard, two-stroke, carbureted, >25 Hp, <75 Hp	20
5	Outboard, two-stroke, carbureted, □75 Hp	11
6	Outboard, two-stroke, direct injected	1
7	Outboard, four-stroke	1
8	Inboard, four-stroke, including stern drive	0.2

Fractions of MTBE/BTEX in Gasoline

To calculate MTBE and BTEX loads, it was necessary to estimate the fraction of each compound in gasoline. A summary of the volume fractions used in modeling is presented in Table 4.2. The sources include the Lake Tahoe motorized watercraft study (Fiore et al. 1998), a U.S. Coast Guard analysis of a reformulated gasoline (Frysinger et al. 1999), and reported values for a California Phase II RFG containing ethanol (Heerman and Powers 1998).

Table 4.2

Fractions of VOC constituents in fuel

VOC	Volume Fraction in Fuel
MTBE	0.140
Benzene	0.010
Toluene	0.072
Ethylbenzene	0.028
Xylene	0.070

Analysis of Boating Census Data

The goal of the boating census was to make reasonable estimates of the daily amount of MTBE entering Donner Lake. Our approach was to develop average MTBE loads and then condition the averages with boating activity information for a particular day. Initially, it was thought that the boating census information would be adequate to directly calculate MTBE loads, but activity survey and fuel use data were too sparse for such a detailed approach.

As a first step, a boating census database was created, consisting of two tables. One contained boat ramp data, the other boating activity survey information. The data were reformatted and consolidated, and records with incomplete data were eliminated. Several fields, including month, day of the week, a weekday/weekend indicator and engine type, were added. The data were broken down by month, by weekend/weekday, by type of activity, and by engine type. Each day was divided into 7 two-hour timeslots, centered on the even hours between 8:00 a.m. and 8:00 p.m.

Assumptions

In approaching the boating data, several initial assumptions were made:

- In boating activity surveys, all boats are counted, and counted only once. (Because of its movement relative to the surveyor, a boat may be counted more than once, or it may not be counted at all.)
- The distribution of boat and engine types on the lake is assumed to be the same as seen at the ramp.
- With respect to boating activity surveys, activity levels in the timeslots that were measured are representative of the whole day.
- Some boats were launched and recovered more than once in a day. For the purpose of the database, an individual boat was allowed only one launch/recovery per day. If there was fuel use information for one of the recoveries, that data record was preferentially retained.

Calculation of MTBE Loading

The first step in calculation of MTBE/BTEX loading was to use the boating activity database to derive average levels of intensity for each type of boating activity, for each month of the boating season, for both weekends and weekdays. The values obtained are presented in Table 4.3.

Table 4.3

Average daily level of boating activity, Donner Lake 2000

	Number of boats - weekend			Number of boats - weekday		
Month	Troll	Ski	Cruise	Troll	Ski	Cruise
June	10	24	37	11	16	32
July	16	37	59	13	19	37
August	14	45	58	9	27	39
September	10	2	7	1	2	3

Then, from the launching ramp database, we obtained the distribution of engine types, the duration of each type of boating activity, and the fuel consumption rate for each type of engine. These data, together with engine emission rates and the volume fraction of the given VOC in gasoline, were applied to the boating intensity data to arrive at average loading. Tables 4.4 through 4.6 show engine distribution, activity duration, and fuel consumption rates, respectively.

Table 4.4

Percentage of engine types, Donner Lake 2000

Engine Types	Weekend			Weekday		
	Troll	Ski	Cruise	Troll	Ski	Cruise
PWC, two-stroke carbureted	0.0	2.4	6.8	0.0	0.0	9.3
PWC, direct injection	0.0	0.0	4.5	5.3	2.3	11.3
OB, two-stroke carbureted, <25 Hp	46.1	1.2	2.3	22.8	4.5	5.2
OB, two-stroke carbureted, 25–75 Hp	15.8	1.2	3.0	15.8	0.0	4.1
OB, two-stroke carbureted, >75 Hp	6.6	7.2	5.3	14.0	11.4	6.2
OB, two-stroke, direct injection	3.9	2.4	0.8	0.0	0.0	0.0
OB, four-stroke	10.5	0.0	0.8	17.5	0.0	4.1
Inboard, four-stroke	17.1	85.5	76.7	24.6	81.8	59.8

Table 4.5

Average activity duration, Donner Lake 2000

Activity Duration by Engine Type (hours)			
Type	Troll	Ski	Cruise
PWC, two-stroke carbureted	0.0	3.5	3.1
PWC, direct injection	0.0	3.0	3.3
OB, two-stroke carbureted, <25 Hp	3.9	2.0	2.8

(continued)

Table 4.5 (continued)

Activity Duration by Engine Type (hours)			
OB, two-stroke carbureted, 25–75 Hp	3.4	2.4	2.7
OB, two-stroke carbureted, >75 Hp	4.2	2.6	2.8
OB, two-stroke, direct injection	4.0	2.4	2.8
OB, four-stroke	3.4	2.4	2.6
Inboard, four-stroke	4.0	2.7	2.6

Table 4.6

Fuel consumption rates, Donner Lake 2000

Fuel Consumption by Engine Type (gallons/hour)			
Type	Troll	Ski	Cruise
PWC, two-stroke carbureted	0*	3	2.17
PWC, direct injection	0*	3	1.87
OB, two-stroke carbureted, <25 Hp	0.32	1.2	0.7
OB, two-stroke carbureted, 25–75 Hp	0.46	1.6	0.67
OB, two-stroke carbureted, >75 Hp	1.31	2.63	1.36
OB, two-stroke, direct injection	0.48	2.4	1.2
OB, four-stroke	0.44	2	0.73
Inboard, four-stroke	0.88	1.88	1.47

* No data for trolling with PWCs

The calculations were performed in a spreadsheet that served as a front end for DLM-MTBE, and a table containing average MTBE loads was exported to a file for input to the model.

Daily MTBE loads were calculated within the model. For days on which no boating activity data were available, the average MTBE load for that month and time of week was used. For days on which whole-lake activity surveys had been conducted, the level of activity was compared to the average level of activity, and the ratio of the two was used as a multiplier by which to condition the average MTBE load.

DLM-2002 CALIBRATION

The object of the calibration process is to ensure that the model accurately represents reality in known situations so that it may be used with confidence as a predictive tool. The hydrodynamic model and the volatilization sub-model do not require calibration for a particular modeling situation. The coefficients used in the model equations are derived from the work of many researchers, and are known to apply under a wide variety of conditions. Degradation processes, however, have not been as well quantified, and there is a range of choice for equation coefficients, as described below.

Hydrodynamic Calibration

Calibration took place in two stages. The hydrodynamic portion of the model was calibrated first by comparing modeled temperature profiles at different points during the season with measured profiles. Through this process it was found that measured wind values had to be multiplied by 1.2 to accurately model the measured lake temperatures. No other adjustments were required. That measured wind values were lower than required is not surprising, considering that the meteorological station was located on the lake shore, where wind speeds would likely be lower than in mid-lake due to shoreline boundary effects.

A comparison of measured and modeled temperature profiles, using adjusted wind values, for three dates during the boating season, is presented in Figure 4.1.

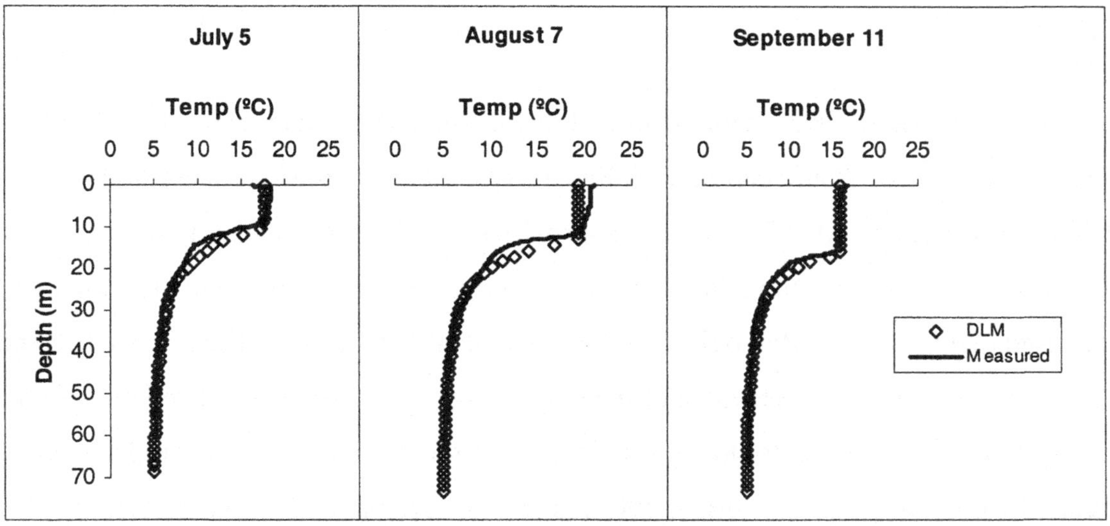

Figure 4.1 Modeled temperature vs. measured temperature for wind multiplier = 1.2

VOC Calibration

Presence of MTBE

The first step in calibrating the MTBE portion of the model was to estimate what fraction of the fuel used at Donner Lake contained MTBE. The fuel sold by the power boat rental concessionaire in 2000 was entirely MTBE-free. Fuel used by these boats was estimated to be ten percent of the total fuel used. Based on fueling location data, 5 percent of the fuel was estimated to come from Nevada, and all of that fuel was assumed to be MTBE-free. All gas purchased from California sources was assumed to contain MTBE. Thus, 85 percent of the fuel used at Donner Lake in 2000 was assumed to contain MTBE.

Measured MTBE Concentration Data

Epilimnetic MTBE was sampled at 2 meters and 6 meters below the surface throughout the investigation. For purposes of model calibration, the two values were averaged to obtain mean epilimnetic MTBE concentration on the sampling dates.

MTBE Calibration

The first calibration run was made without degradation in the model, and using MTBE loading exactly as calculated from boating data. Results of calibration runs are presented in Figure 4.2. The peak modeled MTBE level was approximately 1.7 times the maximum measured value, and occurred later in time. Next, the degradation sub-model was implemented, using rate coefficients based on the work of Rathbun (1998), as described in Chapter 2.The rate coefficients are published as a range, and the high value of the given range was chosen. A second model run was made, and results were very close to values obtained through field sampling. Model results follow measured values very closely through mid-August, then remain 0.2 to 0.8 μg/L higher for the remainder of the boating season.

BTEX Calibration

BTEX calibration was more problematic than MTBE, since there were very few detections above laboratory reporting limits. Toluene was the only one of the compounds that was detected more than once at sampling station 4, occurring several times at concentrations between 0.2 and 0.5 μg/l. The attempt to model toluene forced development of the degradation sub-model, when it became clear that volatilization alone was inadequate to account for the relatively low levels of toluene detected in Donner Lake. The initial run for toluene was made with no degradation factor, and the result is shown in Figure 4.3. Modeled levels are up to 5 times higher than measured concentrations. In the interest of consistency in the modeling process, the high limit of the range of rate coefficients was also chosen for modeling toluene. As with MTBE, a second run was then made, using the degradation sub-model. The result is plotted in Figure 4.3.

Estimated toluene concentrations were very close to measured values for the three toluene samples above detection limits.

There were no detections of benzene and ethylbenzene by which to calibrate the model. However, when volatilization and degradation were applied in the same manner as for MTBE and toluene, modeled concentrations of both compounds remained below detection limits. Figure 4.4 presents model estimates for benzene and ethylbenzene.

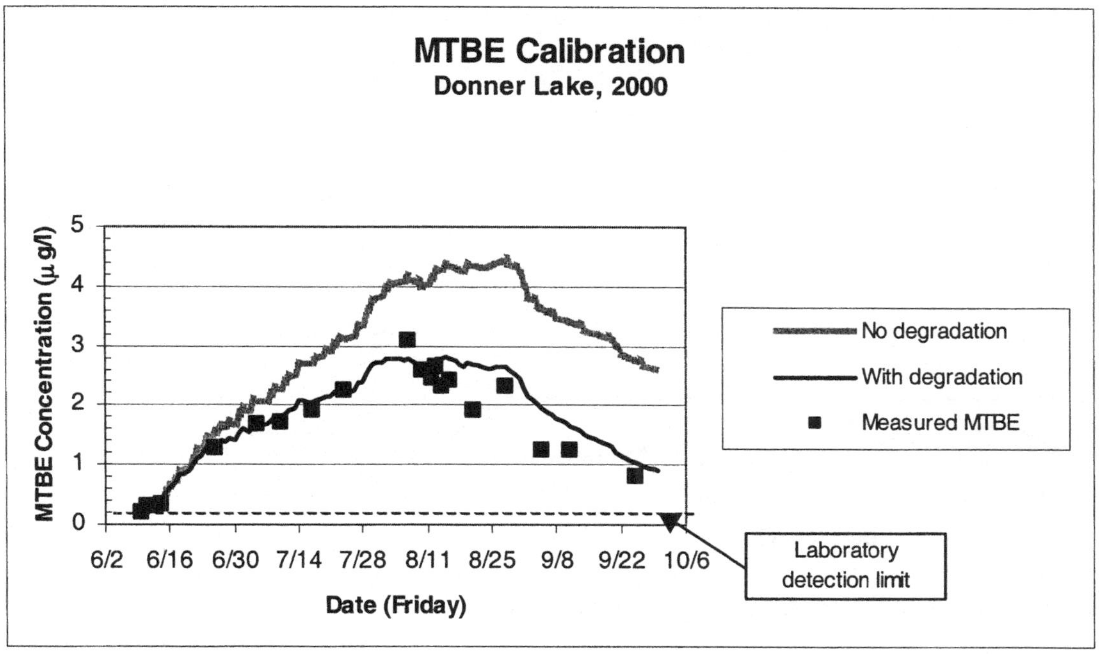

Figure 4.2 MTBE calibration runs

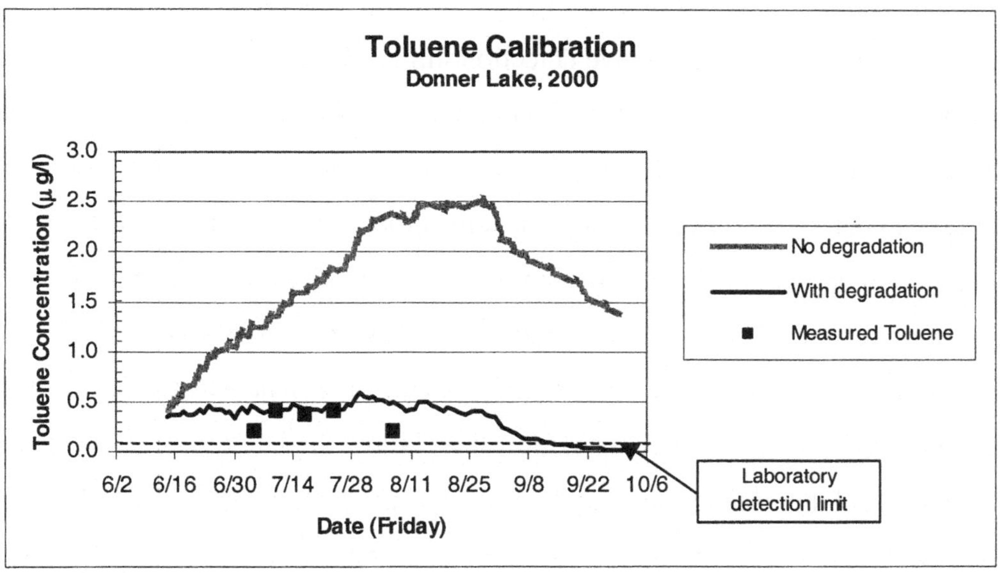

Figure 4.3 Calibration model runs for toluene

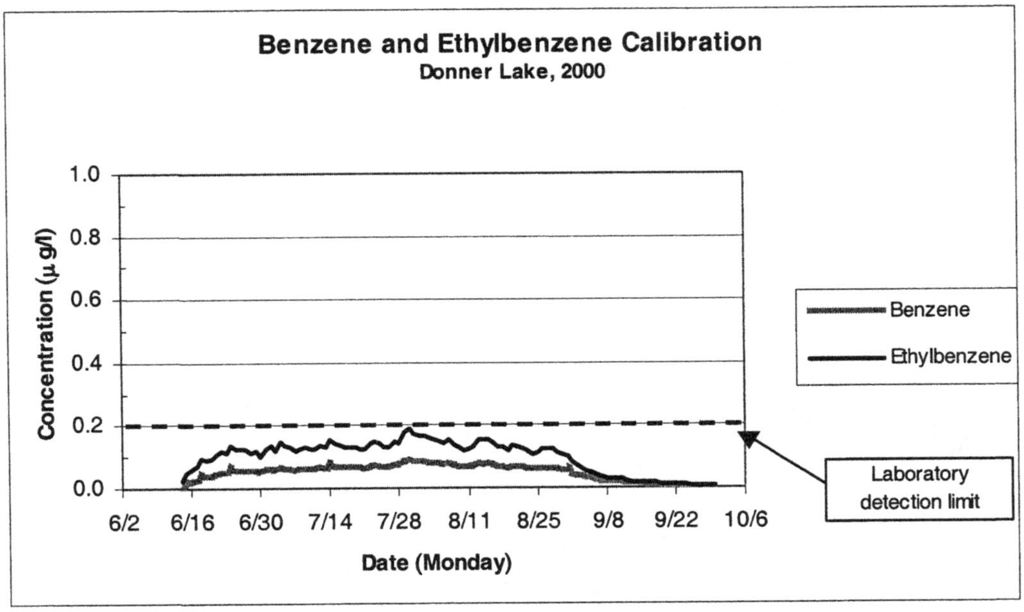

Figure 4.4 Calibration runs for benzene and ethylbenzene

The three isomers of xylene were treated as a single compound for modeling purposes. As with benzene and ethylbenzene, there were no measured values with which to compare model results. The results of calibration runs are shown in Figure 4.5. Even with the degradation sub-model, xylene concentrations remained well above detection limits. Xylene is similar to toluene in its volume fraction in gasoline and in its volatilization rate, as a comparison of Figures 4.5 and 4.3 shows. Its degradation rate, however, is 28 percent lower. Since xylene remained below detection limits during the boating season, it is clear that some other mechanism that reduced xylene concentrations was in operation. Engine emission rates are based on the behavior of MTBE, and might be quite different for xlyene. An unidentified degradation mechanism might also be at work. Therefore, the only way to "calibrate" DLM-2002 for xylene was to increase the degradation coefficient enough to drive modeled xylene values below the laboratory detection limit of 0.2 μg/L. A seven-fold increase in the "standard" degradation rate was required. Because xylene did not behave consistently with the other VOCs, levels of xylene cannot be modeled with confidence in reservoirs other than Donner Lake.

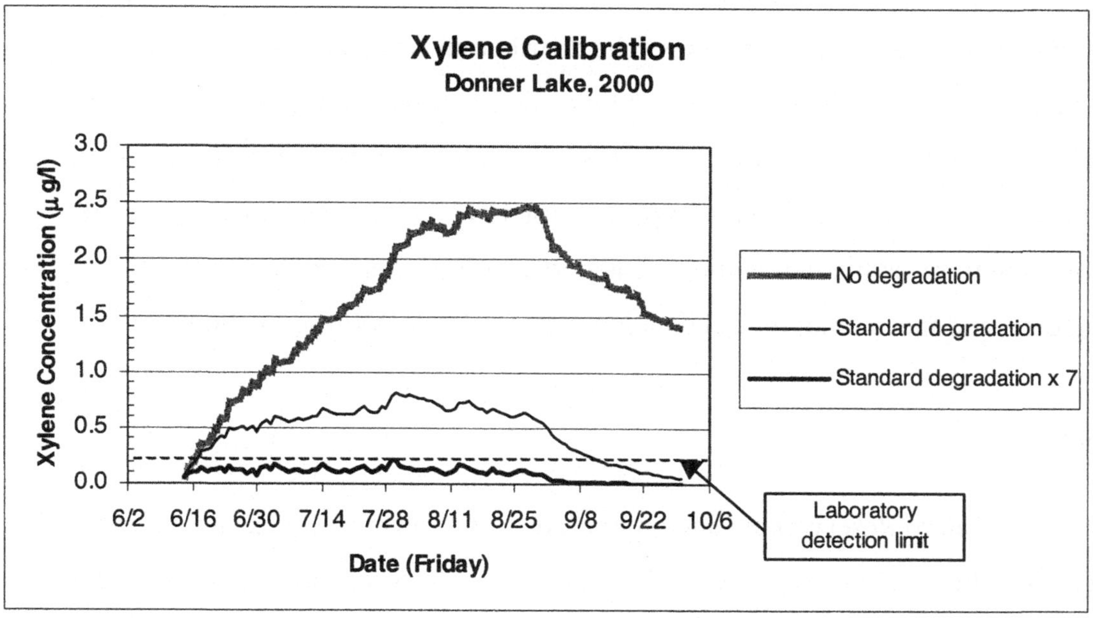

Figure 4.5 Calibration results for xlyene

MODEL VERIFICATION

In order to verify that DLM-2002 is applicable to reservoir situations different from Donner Lake, it was tested on data obtained in 1997 from Calero Reservoir, near San Jose, California, and Lake Perris, near Los Angeles. Calero Reservoir is a relatively small and shallow body of water with an average of 53 boats per day and an average boating density of 37 boats per km^2 during boating season. Boating density is calculated by dividing the total number of boats per day by the reservoir surface area. Not all boats are on the lake at the same time. Lake Perris is located near Riverside, CA, in an arid environment and has an average count of 440 boats per day and a density of 47 boats per km^2. Lake Perris is unique in that its inlet is a submerged pipe rather than a surface stream. A slightly modified inflow algorithm was used to account for the submerged inlet. The model was tested with and without the inflow modifications, and the maximum difference between the two was approximately 0.5 μg/L, or 3.3 percent at times of peak MTBE concentrations.

Boating data for both reservoirs consisted of a daily boat count. To convert boat counts to MTBE load, the boats were subdivided into engine types and activities according to the patterns observed at Donner Lake. The procedure described under "Calculation of MTBE Loading," above, was used to estimate the MTBE load. Since the data were collected in 1997, two changes were made to the assumed boat mix. In the Donner 2000 data, PWC were split 47 and 53 percent between carbureted and fuel-injected engines. The distribution was changed to 95 and 5 percent for carbureted and injected engines, to reflect the prevalence of the conventional-technology two-strokes in 1997. Second, four-stroke outboards constituted 15.6 percent of the outboard population in 2000. This was changed to 3.6 percent for 1997, and the remainder was split evenly between the three carbureted outboard categories. MTBE loads were calculated directly from boating data, and the amount of fuel containing MTBE was estimated at 90 percent.

Calero Reservoir

Results for Calero Reservoir are shown in Figure 4.6. The model picks up the early season measured values and the September 22 result quite well. Anomalous measured values were collected on August 11 and October 27. The measured data points each represent an average of concentrations from three depths within the epilimnion, so it is not a question of a single bad sample. There is, however, nothing in the boating numbers that would imply such low values on those two occasions.

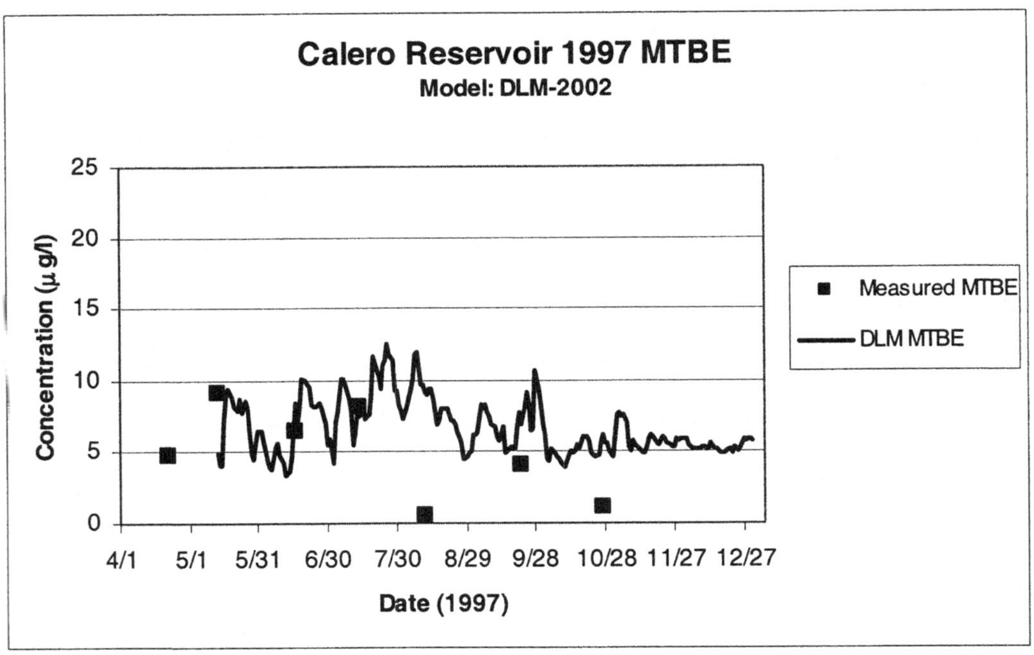

Figure 4.6 Measured vs. modeled MTBE in Calero Reservoir, 1997

Lake Perris

Model results for the 1997 data at Lake Perris are presented in Figure 4.7. The model reasonably approximates most of the measured data except for high MTBE concentrations on August 7 and September 2, 1997. August 7 was a Thursday, with no exceptional boating numbers over the previous week. September 2 was the day after Labor Day in 1997, and while boating had been high the previous 3 days, it was not the highest level of the season.

COMPARISON OF DLM-1998 AND DLM-2002 MODELS

DLM-1998 and DLM-2002 were compared by running both models on the 1997 and 2000 data sets from Donner Lake.

To compare DLM-2002 and DLM-1997 using the 1997 boating data, three model runs were made. First, loading was calculated according to the 1998 algorithm, in which the daily boat count is multiplied by the fraction that are two-strokes, and then by a fixed load of 350,000 mg (.125 gallon) of MTBE per boat. DLM-1998 and DLM-2002 were then run using the daily loading values thus calculated. Loading for DLM-2002 is calculated by breaking the daily boat

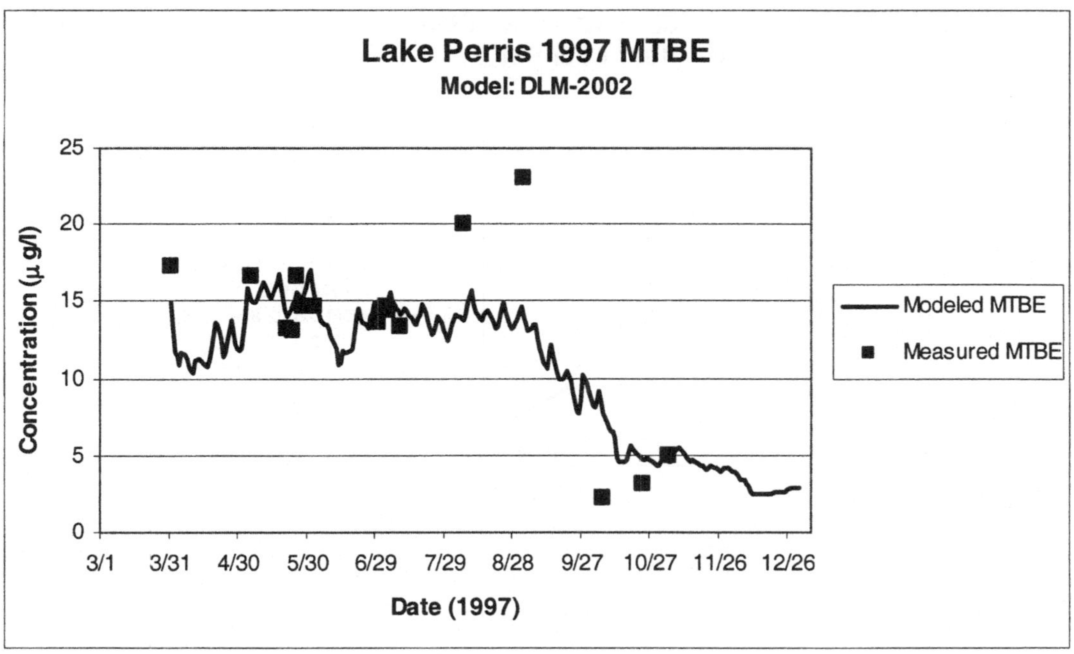

Figure 4.7 Measured vs. modeled MTBE in Lake Perris, 1997

count down into categories of engines and activities and summing the contributions of each category. For the third model run, loading was calculated using the 2002 algorithm on the 1997 boating data. As noted earlier, the engine type distribution was modified for 1997 to account for the greater number of carbureted two-stroke engines. The results for the 1997 boating data are presented in Figure 4.8.

When DLM-2002 is used with loads as calculated by the 1998 method, it follows measured data as well or better than the earlier model. However, when the 2002 loading algorithm is applied to the 1997 data, DLM-2002, produces concentrations that are much lower than those of DLM-1998. The implication is that boating counts in 1997 are quite low relative to actual conditions, and that a high per-boat-loading factor was used to compensate.

Comparisons for the 2000 data set are presented in Figure 4.9. To apply the 1998 loading algorithm to the 2000 boating data, an estimate of the fraction of the 2000 boats that were two-strokes was calculated. The daily boat counts were then multiplied by that fraction (0.3), by the estimated fraction of fuel containing MTBE (0.85), and by the 1998 estimate of loading per two-stroke boat (0.125 gallons of MTBE). When DLM-1998 is run on the resulting daily load, the concentrations of MTBE are over-estimated.

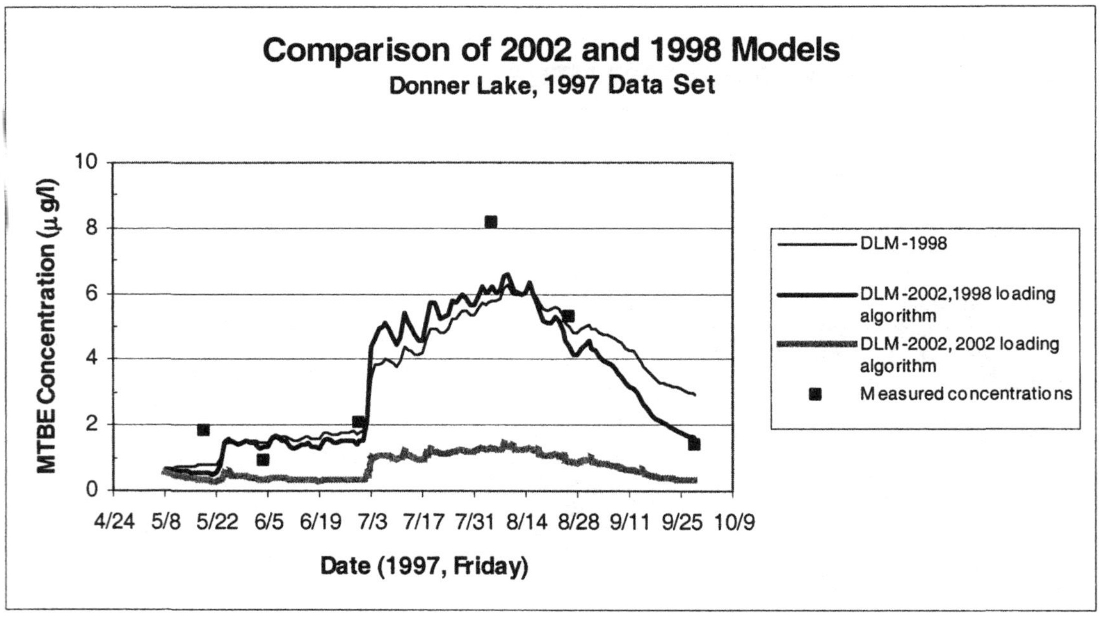

Figure 4.8 Comparison of DLM-2002 and DLM-1998 using 1997 data

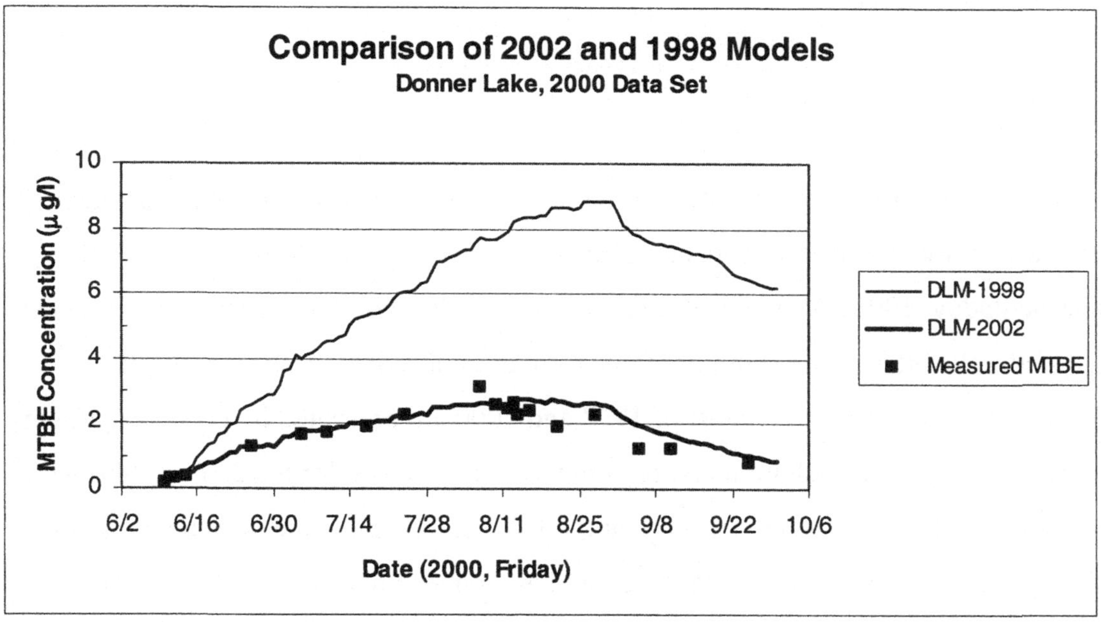

Figure 4.9 Comparison of DLM-2002 and DLM-1998

To determine the source of the differences, a composite per-boat MTBE load was calculated for 2000. The composite load includes the effects of all the different types of engines and activities. The value for Donner Lake in 2000 was 0.027 gallons per boat. This indicates that the per-boat load of 0.125 gallons used for 1997 was quite high. Likewise, the average daily number of boats in July and August of 1997 was 38. For 2000, the number was calculated to be 106. Since it is unlikely that the difference between the two seasons was that great, it seems clear that the number of boats was under-estimated in 1997. The combination of high per-boat loading and low boating numbers allowed DLM-1998 to provide reasonable estimates of MTBE concentrations.

COMPARISON OF 1997 AND 2000 BOATING SEASONS

Number of Boats

As noted above, the average boating numbers for 1997 and 2000 are quite different. The 1997 boat numbers consisted only of launches from the public boat ramp and the Tahoe-Donner Association (private) ramp. Boats that were resident on the lake were apparently not taken into account. In 2000, the whole-lake binocular surveys allowed estimates to be made of the total boating activity on Donner Lake.

Weather

In 1997, weather data were obtained from the Northern Sierra Air Quality Management District. The measurement station was located approximately 1 mile south of the lake outlet (McCord and Schladow 1998). Meteorological data for the 2000 boating season were obtained from a MET station located on the shore of Donner Lake, as described earlier. Average meteorological values for July and August of 1997 and 2000 are presented in Table 4.7. Values for short-wave and long-wave radiation and air temperature are close for the 2 years. Short-wave and long-wave values for 1997 were measured at the UCD Lake Tahoe meteorological station near Tahoe City. For 2000, short wave was measured at Donner Lake, and longwave was calculated using the equation presented by Chapra (1997). Wind, the most significant weather-related variable in the modeling, was much higher in 1997 than 2000. It seems unlikely that 1997 was that much windier, and a substantial portion of the difference is probably due to the different locations at which wind was measured.

Table 4.7

Average meteorological values for July and August

Year	Short-wave radiation (KJ/m2•d)	Long-wave radiation (KJ/m2•d)	Air Temp (°C)	Vapor Pressure (mbar)	Wind (m/sec)
1997	25,554	26,732	16.5	9.6	2.96
2000	24,413	26,244	16.5	7.52	1.83
Percent Difference	-4.5	-1.8	0.0	-21	-38

MTBE Fuel Content

In the modeling for 1997, all fuel was assumed to contain MTBE. The assumption seems reasonable, since the powerboat concessionaire was using fuel with MTBE at that time, and the only suspected source of MTBE-free fuel would be that coming in from Nevada. For the 2000 modeling, data were available which allowed quantification of the amount of MTBE-free fuel at about 15 percent, as outlined earlier.

CHAPTER 5
SPREADSHEET MODEL

INTRODUCTION

In addition to the more complex DLM model, UCD developed a simple spreadsheet program that estimates MTBE/BTEX concentration levels in a lake or reservoir. The user enters information for a local waterbody, chooses a compound, and the program estimates the maximum concentration of the VOC that can be expected during the boating season. The estimator is based on an extensive series of runs with DLM-2002.

This tool allows reservoir managers to quickly and simply determine if MTBE levels in their reservoir have the potential to be problematic. Additionally, the spreadsheet can be used to roughly compare different scenarios in a waterbody. It is important to note that the spreadsheet only delivers a seasonal maximum value. For more detailed information, the full DLM-MTBE model should be used.

METHOD OF ESTIMATION

A "base case" DLM modeling scenario was developed, based on weather and hydrologic conditions encountered at Donner Lake in 2000. The base case included an air temperature of 20ºC, short-wave radiation at 24,000 KJ/m^2•day, clouds covering 33 percent of the sky (for calculation of longwave radiation), vapor pressure of 10 millibars, and a wind velocity of 2 meters per second. Inflow and outflow were held constant. The DLM model was run for 93 days under these conditions (nominally from June 14 to September 15).

The first set of DLM runs were made with varying specific loads (load per unit area) of MTBE and BTEX, to determine the highest concentration levels reached for each case. A set of curves for VOC loading versus the resulting VOC concentration in the lake was developed, and these are used in the estimator to establish the base case concentration for the contaminant of interest. Then, a set of DLM runs was made for each environmental variable, varying it over the expected range of values. From these runs, regression curves were developed which related changes in the input variable to changes in the VOC concentration in the water.

The spreadsheet user enters average daily meteorological information, basic reservoir physical parameters, and boating use data. The VOC level for the user's case is estimated by first computing the specific VOC loading and calculating the resultant VOC level in the reservoir. Then, for each meteorological variable, the difference between the user's case and the base case is determined and the appropriate regression curve is used to calculate an adjustment to the VOC concentration.

One limitation of the estimator is that if it is used for a shallow reservoir which mixes to the bottom, the estimated VOC level may be low. The estimator calculates the average reservoir depth and the expected maximum mixing depth, and issues a warning if they are within 4 meters of each other.

TESTING

The estimator was initially tested for the Donner Lake 2000 case on which it was based. Measured values of MTBE, DLM daily estimates of MTBE, and the estimator value are presented in Figure 5.1. Figures 5.2 through 5.5 present the same quantities for benzene, toluene, ethylbenzene, and xylene. Note that estimator values for xylene should be used with caution, since xylene required special treatment in DLM calibration. The estimator produced appropriate maximum values for MTBE and the BTEX compounds in Donner Lake. In each case, estimator values were slightly above the maximum level predicted by DLM. For MTBE, the estimator value was slightly lower than the maximum measured value (2.99 ppb vs. 3.10 ppb, respectively). With respect to potential water quality impacts, these two values are not significantly different.

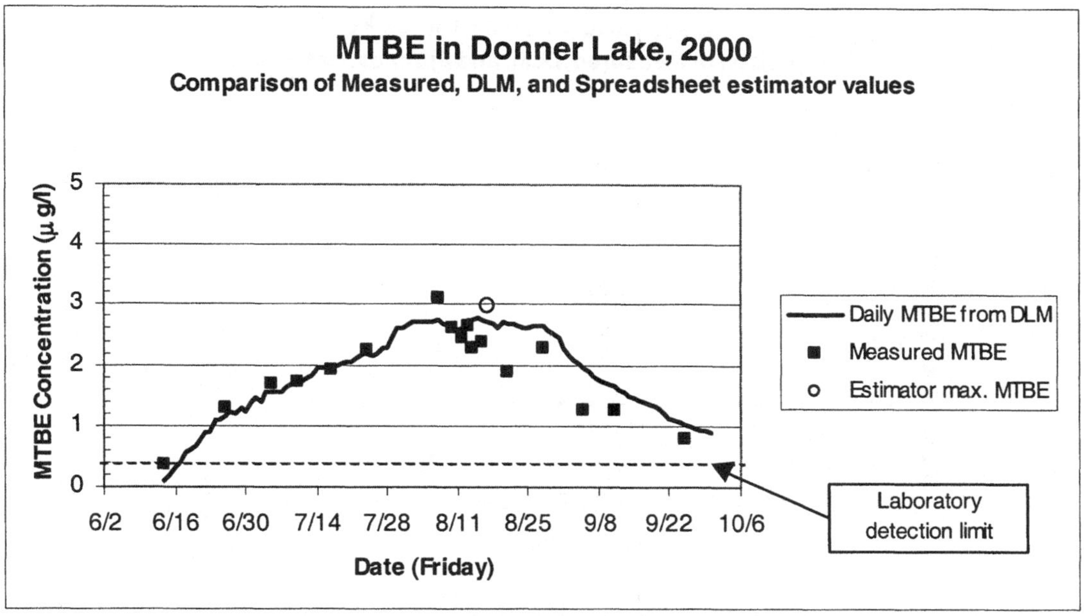

Figure 5.1 Spreadsheet estimator vs. DLM for MTBE in Donner Lake

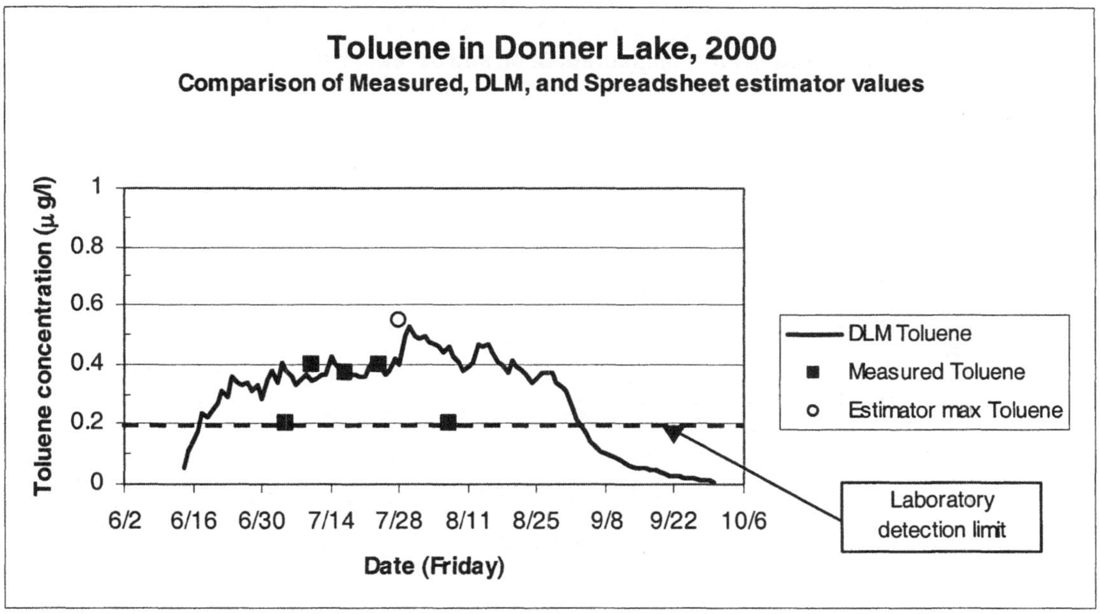

Figure 5.2 Spreadsheet estimator vs. DLM for Toluene in Donner Lake

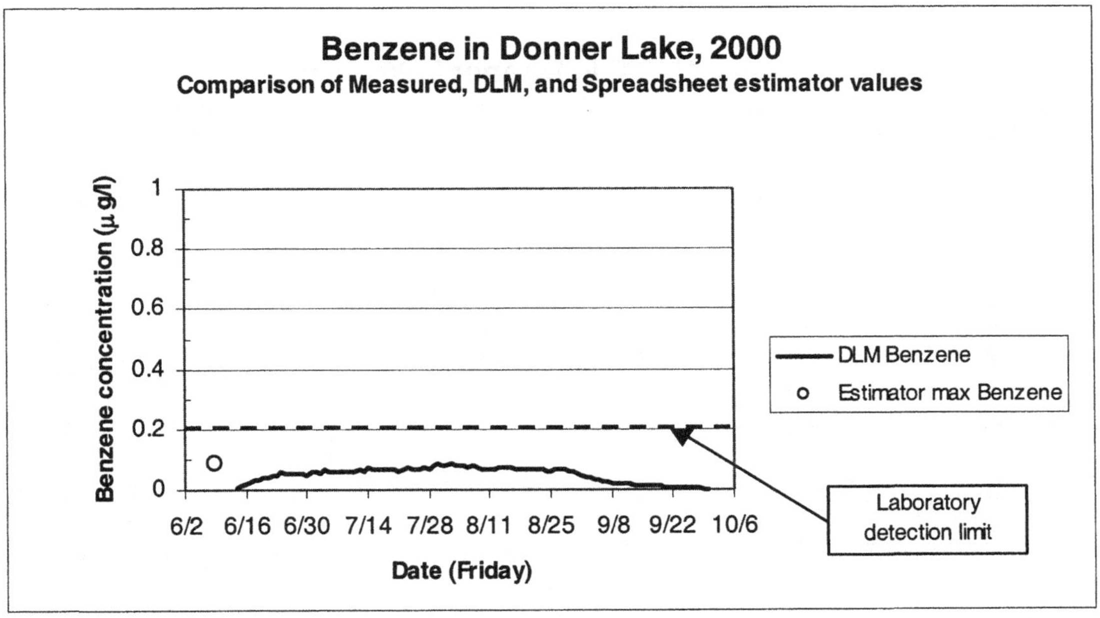

Figure 5.3 Spreadsheet estimator vs. DLM for Benzene in Donner Lake

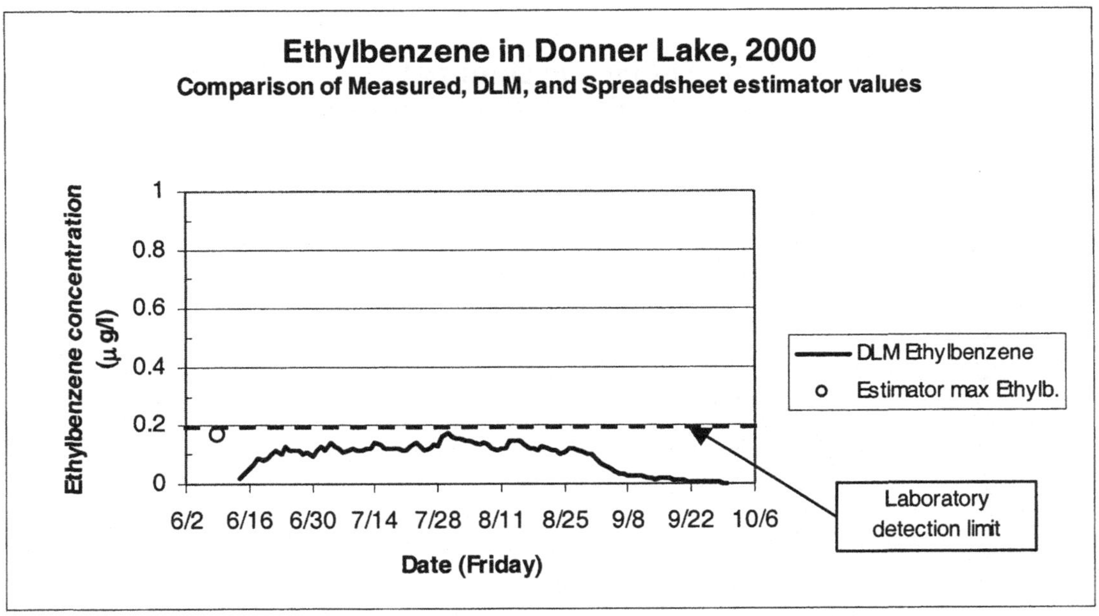

Figure 5.4 Spreadsheet estimator vs. DLM for Ethylbenzene in Donner Lake

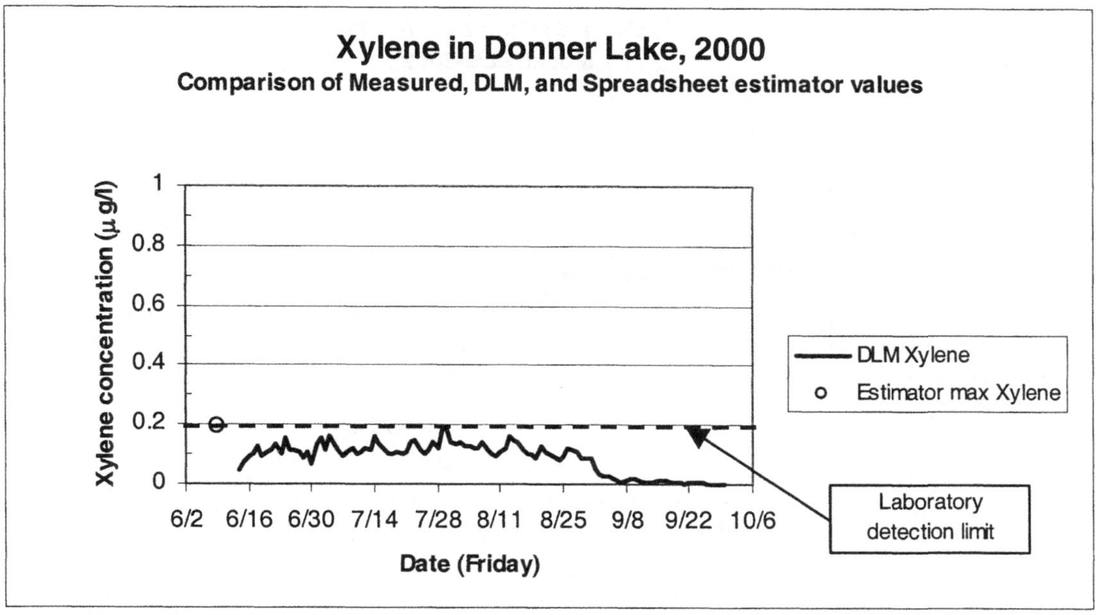

Figure 5.5 Spreadsheet estimator vs. DLM for Xylene in Donner Lake

DLM and the estimator were then applied to 1997 data from Calero Reservoir (Santa Clara Valley Water District) and Lake Perris (Metropolitan Water District). Since the data were taken in 1997, two adjustments were made to the estimator. In the Donner 2000 data, PWCs were split 47 and 53 percent between carbureted and fuel-injected engines. The distribution was changed to 95 and 5 percent for carbureted and injected engines, to reflect the prevalence of the conventional-technology two-strokes in 1997. Second, at Donner, four-stroke outboards constituted 15.6 percent of the outboard population. This was changed to 3.6 percent for 1997, and the remainder was split evenly between the three carbureted outboard categories.

Results for MTBE in Lake Perris and Calero Reservoir are presented in Figures 5.6 and 5.7.

The estimator results for Lake Perris and Calero Reservoir are comparable to those from DLM. For Lake Perris, the estimator maximum MTBE is slightly less than the maximum predicted by DLM (16.7 and 17.0 mg/L, respectively). At Calero Reservoir, the estimator, for 1997, predicts a maximum MTBE level 39 percent lower than the maximum value for DLM. Calero is a relatively shallow reservoir, and physical data show that it did indeed mix fully during the 1997 boating season. The estimator appropriately flags that possibility and indicates that its estimate may be low.

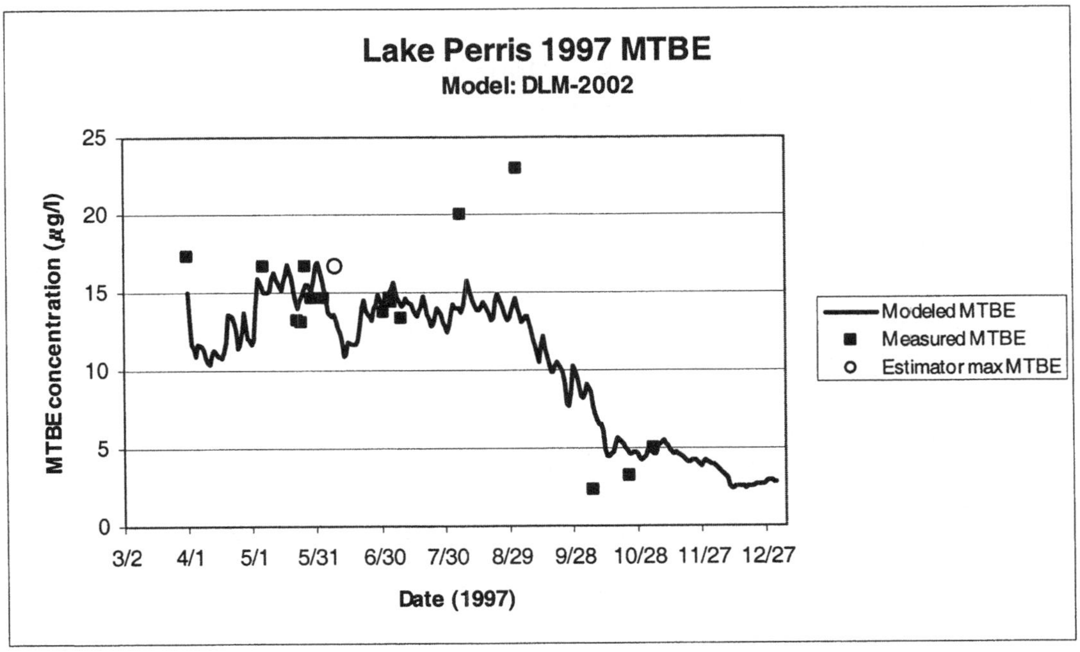

Figure 5.6 Estimator vs. DLM for MTBE in Lake Perris, 1997

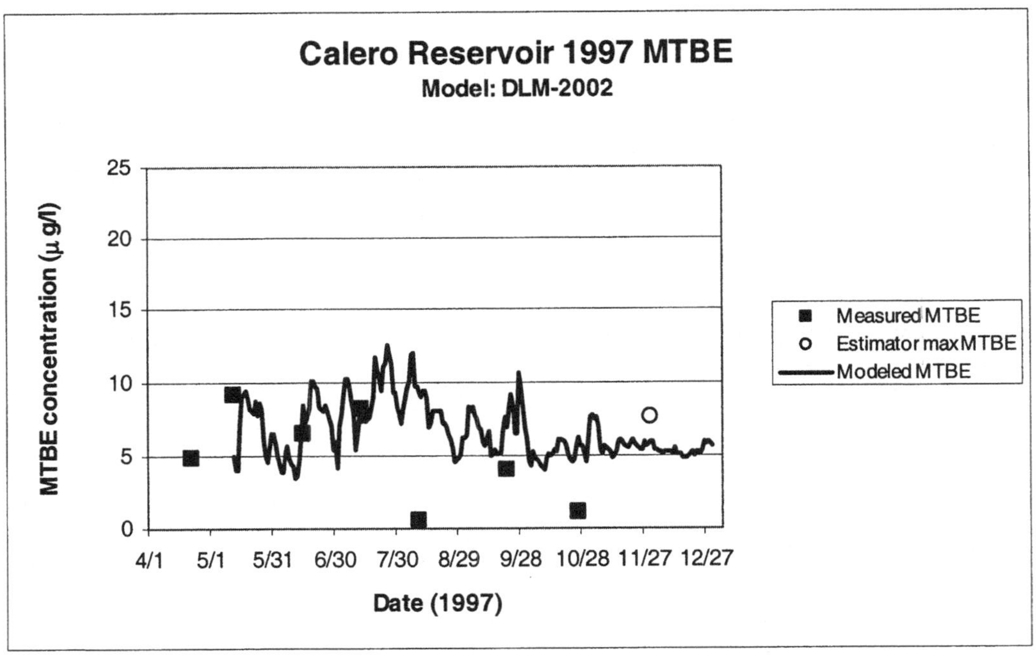

Figure 5.7 Estimator vs. DLM for MTBE in Calero Reservoir, 1997

ESTIMATOR PORTABILITY

Since the estimator was developed based on data taken in California, the question arises whether it is applicable in other regions. The main regional differences would be climate and water clarity. Donner Lake is a relatively clear reservoir located in a high mountain climate. As noted previously, DLM, and by extension the estimator, account for the major meteorological parameters affecting reservoir dynamics. The model and estimator tests conducted with Calero Reservoir and Lake Perris data represent quite different climatic regimes and demonstrate the viability of the estimator under those warmer and drier conditions. Water clarity is represented in the model by a light extinction coefficient that determines how deeply light penetrates the water. An extinction coefficient of 0.24 was used for modeling Donner Lake, a value that is appropriate for Secchi depths of 7-12 meters. DLM modeling of Lake Perris and Calero Reservoir was conducted using extinction coefficients of 0.34 and 1.0, respectively. The estimator, using the Donner Lake extinction value, produced very acceptable results for those two reservoirs. To determine the effect of more turbid water on model results, DLM runs were made with Donner Lake data, using extinction coefficients of 0.5 and 1.0, which correspond to Secchi depths in the 3-6 and 1-2 meter ranges. The results are presented in Figure 5.8. Doubling the extinction coefficient results in an approximately 20 percent increase in peak modeled MTBE level. The reason is that with more turbid water, sunlight does not penetrate as deeply in the reservoir, and the upper mixed layer becomes shallower, resulting in increased MTBE concentration in the mixed layer. At a coefficient of 0.5, the estimator value is slightly lower than the peak DLM-2002 value. The effect increases as the water becomes more turbid. At Secchi depths of 1 to 2 meters (extinction coefficient = 1.0), the difference between the estimator and the DLM value is approximately 25 percent. While not large, the difference is substantial, and future versions of the estimator will include a user estimate of Secchi depth.

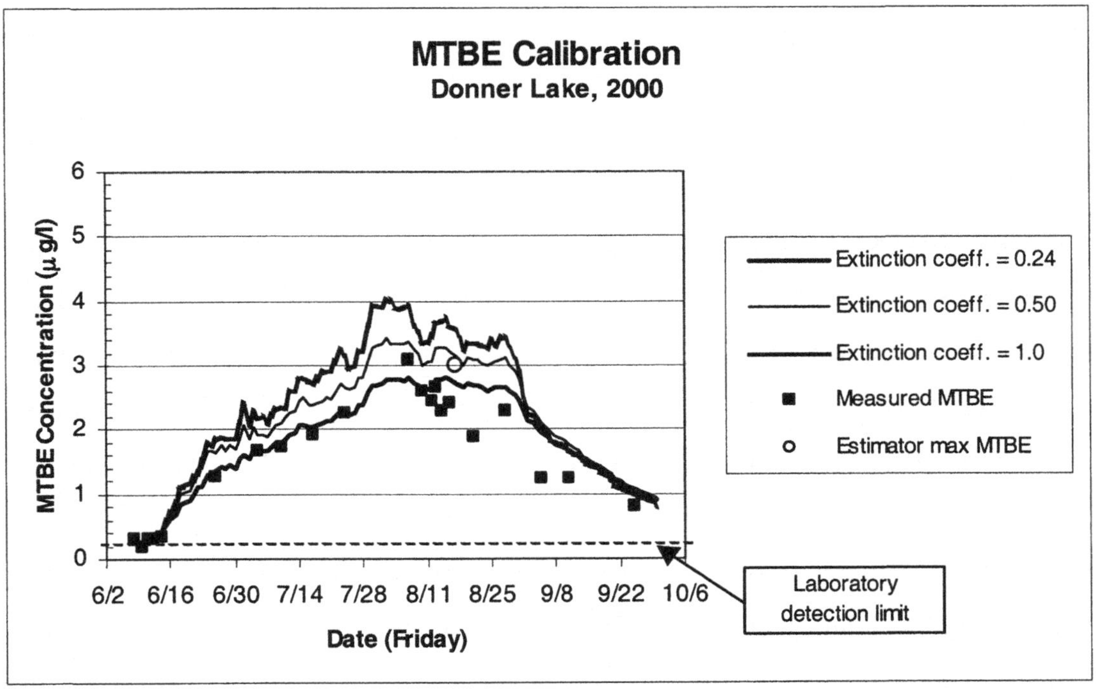

Figure 5.8 Effect of turbidity on MTBE concentration

INSTRUCTIONS FOR USE

A short user guide has been prepared, and is presented in Appendix I. The guide provides basic information about the estimator and gives additional information for each user entry. The user guide is also contained on a worksheet within the spreadsheet estimator itself.

DISCLAIMERS

The estimator includes a note on the main page to the effect that use of the software constitutes agreement to a set of terms and conditions for use. The terms and conditions are detailed on a "Terms and Conditions" page within the estimator. The essential message is that the user is entirely responsible for any results obtained, and agrees to release the UCD from any liability. The conditions are listed below.

Terms and Conditions

By using this software, you consent to the following stipulations and conditions:

1. This software is for rough estimation only. It is intended to be as accurate as its input data allow, but it can provide only a crude representation of any real reservoir situation. There are many conditions that affect VOC levels in reservoirs, which cannot be represented in this estimator.
2. UCD makes no claims or warranties, express or implied, concerning the accuracy or validity of this estimator.
3. The user is entirely responsible for the use of any results obtained from this software. The user agrees to hold UCD and its faculty, staff, and students harmless from any claims resulting from the use of this software, regardless of whether such claims arise from negligence on the part of UCD. The user recognizes that the software has not been fully tested, and may contain errors that affect the accuracy of the results.

CONCLUSIONS

The spreadsheet estimator has been shown to produce reasonable estimates for maximum concentration levels of MTBE and BTEX. It has been tested for several different reservoir types and climates. Within its stated limitations (lakes that do not mix to the bottom), it produces values consistent with DLM predictions. In one case tested, its estimate was substantially lower than the highest measured concentration. In general, however, the estimator should provide managers with an estimation accurate enough for decision making. This spreadsheet estimator should prove to be a useful tool by which reservoir operators can estimate the magnitude of potential VOC contamination under various management options.

CHAPTER 6
RESERVOIR MANAGEMENT OPTIONS

INTRODUCTION

Several options can be employed to manage MTBE and BTEX contamination in reservoirs. The feasibility of each option (e.g., political and public acceptance, cost, operation and maintenance, etc.) must be evaluated to suit the needs of a particular reservoir, audience, and geographic location. Based on the results of this research study, the options proposed for MTBE and BTEX management in water supply reservoirs are categorized into: (A) source control; (B) treatment; and (C) operations management. Source control options through management of recreational boating include: (i) banning recreational boating; (ii) restricting recreational boating; (iii) restricting MTBE content in recreational boating fuel; and (iv) restricting or eliminating on-lake fueling. Treatment options involve using advanced technologies in combination with conventional water treatment processes. Operations management options include changing the location of the drinking water intake in the reservoir and changing the timing of drinking water withdrawal. This chapter presents a detailed description of the options proposed to manage MTBE and BTEX in reservoirs and the methodology and assumptions used to understand and evaluate the various options. It also describes the economic factors to consider when choosing between the management options.

Description of Methodology and Assumptions

The management options were evaluated using DLM-2002 and data obtained at Donner Lake between June and September 2000. Model output for MTBE and BTEX in the upper mixed layer of the reservoir (i.e., the epilimnion) was taken at a depth of 4 meters. Since the Donner Lake epilimnion was 8 meters deep or greater at all times during the boating season, and since model output was taken at midnight when the epilimnion was well-mixed, the 4-meter depth can be considered representative of the epilimnion as a whole. Three boating densities were defined for management scenario runs: (1) medium (the actual number of boats recorded at Donner Lake); (2) low (half the actual number of boats); and (3) high (twice the actual number of boats).

DESCRIPTION OF MANAGEMENT OPTIONS

Source Control through Management of Recreational Boating

As discussed in Chapters 1 and 2, MTBE contamination occurs when unburned fuel is released into the water. As stated above, source control through management of recreational boating can be achieved by: (1) banning recreational boating; (2) restricting recreational boating; (3) restricting the content of MTBE in recreational boating fuel; or (4) restricting or eliminating on-lake fueling.

Banning Recreational Boating

To evaluate the rate at which MTBE would disappear from a lake following the cessation of recreational boating, three scenarios were modeled: (i) unrestricted boating throughout the boating season (ii) a ban on boating starting August 1, 2000 and (iii) a ban on boating starting August 15, 2000. The boating density was assumed to be medium, that is, the actual boating density encountered at Donner Lake. These model runs are depicted in Figure 6.1.

When boating was unrestricted, predicted MTBE concentrations peaked on August 15 at 2.8 ppb, and then fell steadily, dropping below 1 ppb (as a basis for comparison) on September 26, after 42 days. When recreational boating is banned after the August 15 peak, the 1-ppb level is reached on September 3, after 19 days. For the August 1 ban scenario, peak MTBE level was capped at 2.5 ppb and the 1-ppb level was reached 18 days later, on August 19. The August 1 and August 15 bans are similar in their starting concentrations and the amount of time required to reach a given MTBE concentration. With unrestricted boating, however, it takes much longer for MTBE levels to be reduced, because of both the continued MTBE inputs and the cooler weather in September, which slows down the volatilization and degradation processes.

MTBE concentrations on September 30, at the end of the boating season, are 0.86 ppb, 0.35 ppb, and 0.16 ppb for the unrestricted, August 15 ban and August 1 ban, respectively. In summary, banning recreational boating is highly effective in reducing MTBE concentrations. To achieve primary and secondary CA MCLs, banning recreational boating starting August 1 or August 15 appears to be effective with the former resulting in lower concentrations.

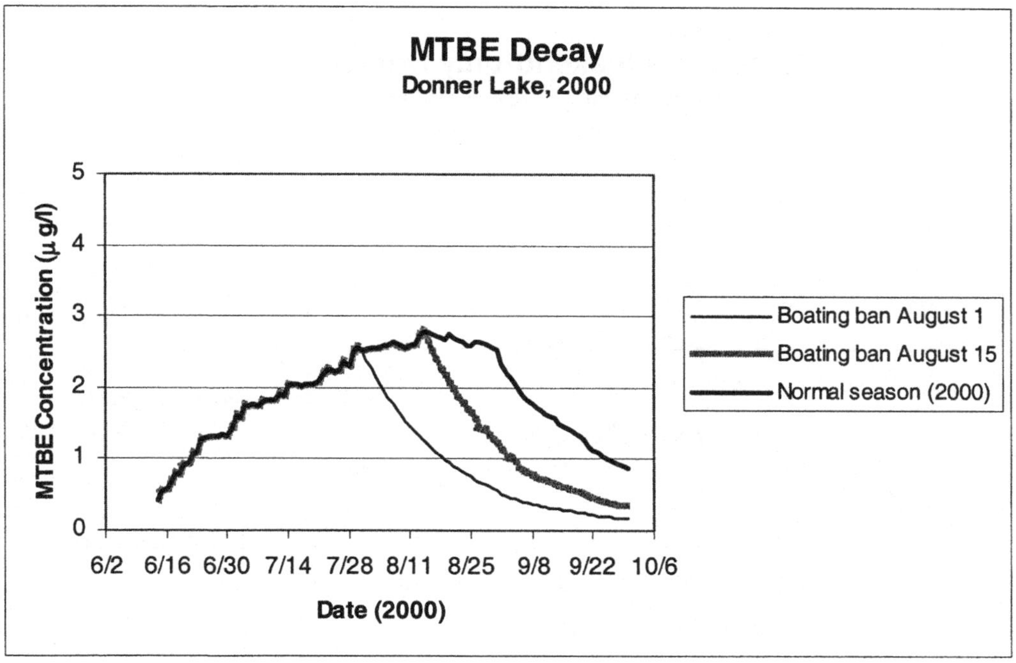

Figure 6.1 MTBE decay for three scenarios

The effect on MTBE concentration of banning boating (on either weekends or weekdays) was also investigated. Those management options were run using the high-density boating scenario. A comparison of predicted MTBE levels for full-time and weekend-only boating is shown in Figure 6.2. The results for weekday-only boating (not shown) are very similar. Peak seasonal MTBE level is reduced from 5.6 to 3.2 ppb, which for this scenario is the difference between being out of compliance and meeting the CA-MCL. It is clear that restricting recreational boating to certain days of the week can be quite effective in reducing MTBE levels, and may be a viable option when a complete ban is not feasible.

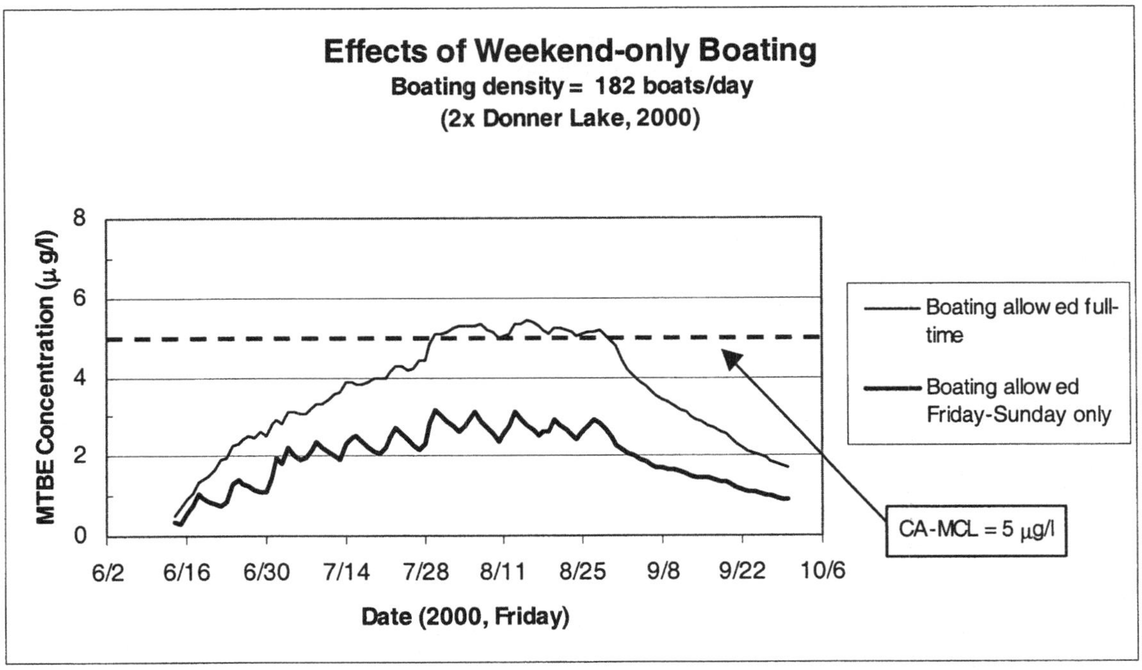

Figure 6.2 Effects of weekend-only boating

Restricting Recreational Boating

Management options for reducing MTBE concentrations in reservoirs include restricting the types of engines or the number of boats on the water. The effects of each of these controls were investigated using DLM-2002.

For a given level of boating activity, the mix of engine types is the primary determinant of the amount of fuel, and thus MTBE, released to the water. Two-stroke engines with carburetors have been found to release significantly higher amounts of unburned gasoline than direct-injected two-stroke or four-stroke engines (see Chapter 3). To understand the effect of the different engine types on MTBE concentrations, three scenarios were modeled. The first was the Donner Lake situation as encountered in 2000, with the boating density and engine mix exactly as observed in the boating census. In the second run, all of the older technology (carbureted) two-stroke engines were banned. For the third scenario, all older two-strokes were replaced with advanced technology two-strokes, including direct fuel injection. The results are presented in Figure 6.3. The average boating density encountered at Donner Lake was 91 boats per day, or 0.25 boats/hectare/day. When older technology two-stroke engines are prohibited, the peak MTBE level for the season is 0.3 ppb, an 89 percent reduction from the peak concentration with no engine restrictions. If all two-stroke engines are converted to advanced technology two-strokes, the peak MTBE concentration becomes 0.5 ppb, an 82 percent reduction. If the number of boats is doubled, to 182 boats per day (0.5 boats/hectare/day), the peak concentrations are approximately doubled, and the percent reductions in MTBE remain similar, as shown in Figure 6.4. These results demonstrate the impact of advanced technology two-stroke engines in reducing MTBE concentrations in surface waters, an impact which becomes more pronounced at higher levels of boating activity.

The level of recreational boating, or boating density, is the most important factor governing MTBE and BTEX levels in lakes and reservoirs. To determine the effect of different levels of boating activity, MTBE in Donner Lake was modeled with low, medium, and high boating densities, as previously defined. In addition, when a lake is used for both boating and water supply, it is useful to know what level of boating could cause a violation of water-quality standards. Several model runs were made to determine the level of boating activity that would cause the peak MTBE concentration to reach the California secondary MCL of 5 ppb. The results are presented in Figure 6.5.

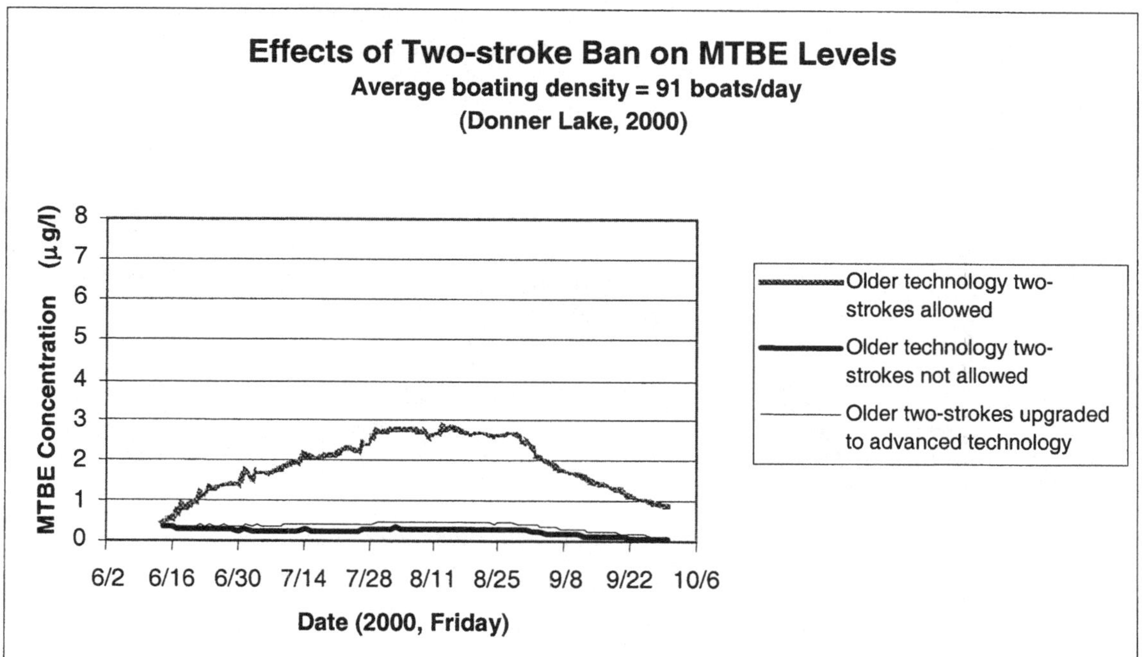

Figure 6.3 Effect of engine mix on MTBE concentrations, medium density boating

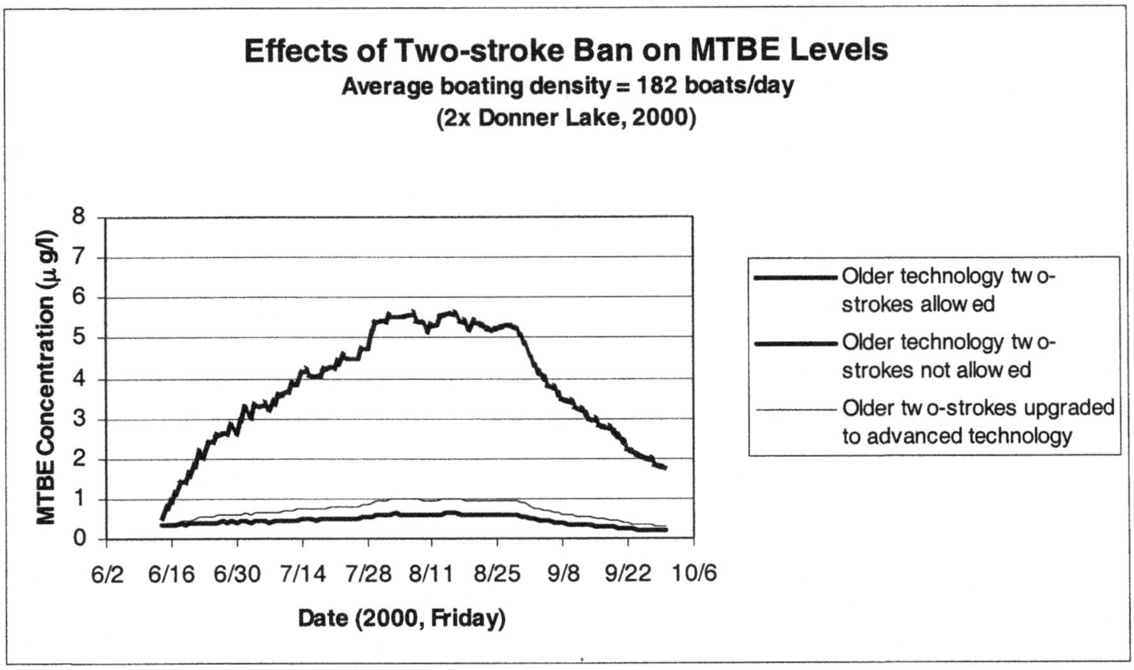

Figure 6.4 Effect of engine mix on MTBE concentrations, high density boating

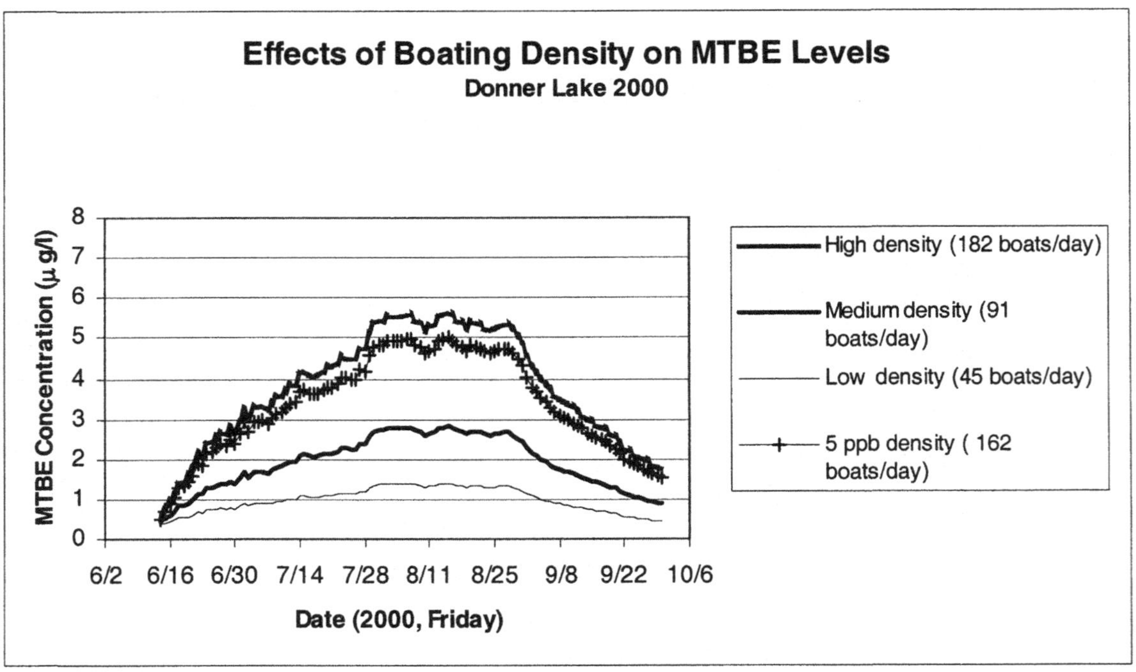

Figure 6.5 Effect of boating density on MTBE level, Donner Lake

The predicted concentrations of MTBE for the varying boating densities indicate that peak concentrations were reached in a similar period of time and that the pattern of MTBE concentrations over the boating season is similar in all cases. The predicted peak concentrations at low, medium and high boating density were 1.4, 2.8 and 5.6 ppb, respectively. An average boating level of 162 boats per day (0.45 boats/ha/day) was required to reach the 5-ppb MTBE level in the epilimnion. These results indicate that MTBE concentration in the lake is directly proportional to boating density.

Figures 6.6, 6.7, and 6.8 demonstrate the effects of varying the density of boats during a given period. A peak weekend was chosen and modeled for low (50 percent), medium (100 percent), and high density (200 percent) boating. The peak concentration for the high-density scenario, 5.27 μg/L, is about four times the peak level from the low boating density scenario of 1.3 μg/L.

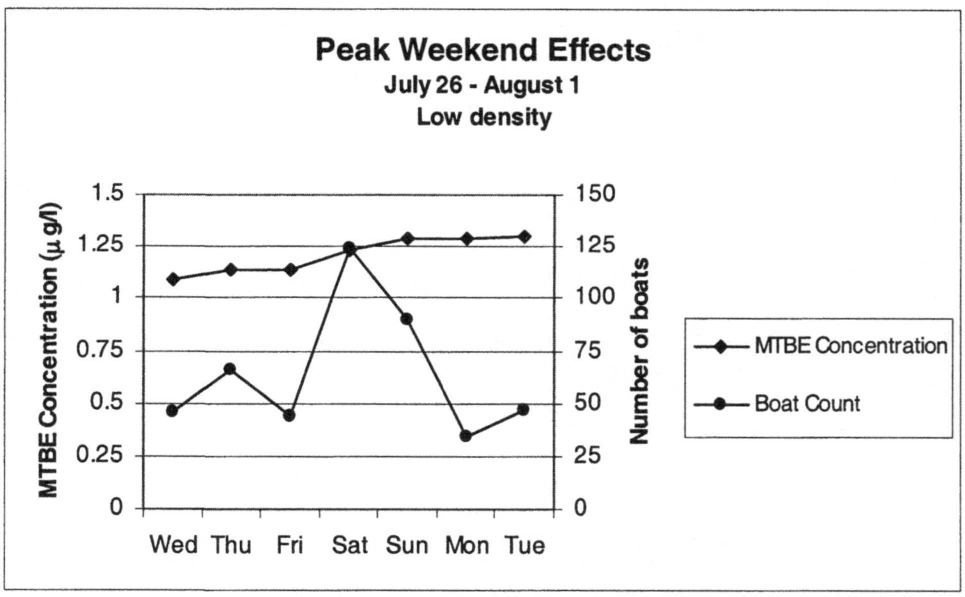

Figure 6.6 Peak weekend effects after low density boating

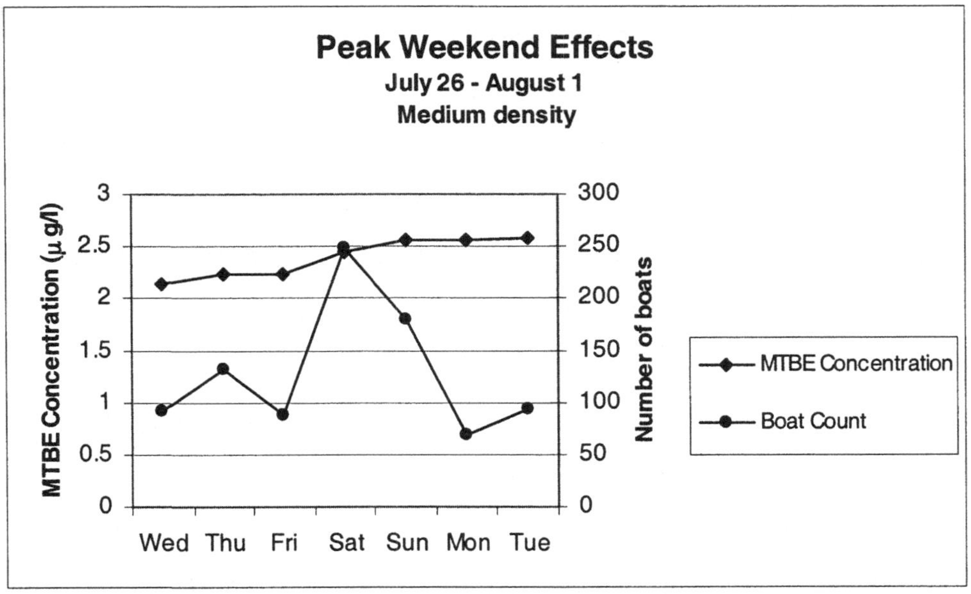

Figure 6.7 Peak weekend effects after medium density boating

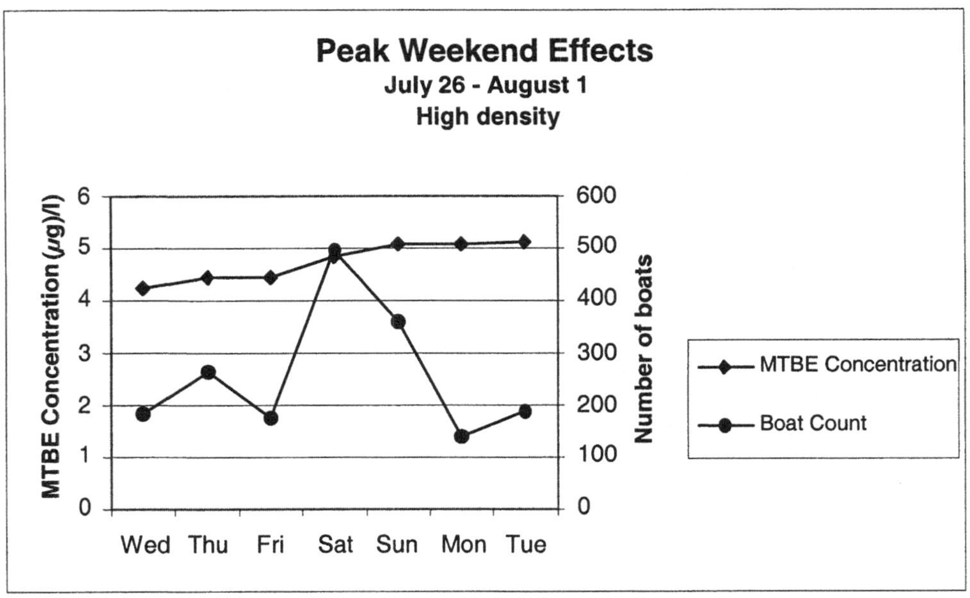

Figure 6.8 Peak weekend effects after high density boating

Restricting MTBE Content in Recreational Boating Fuel

For Donner Lake, it was estimated that 85 percent of the fuel burned at the lake contained MTBE. MTBE content in the fuel supply has a direct effect on the amount of MTBE entering surface waters. Since the regulatory climate is different in other states, and is changing in

California, that percentage is likely to vary considerably depending on the reservoir situation. To assess the impact of varying amounts of MTBE in the fuel supply, model runs were generated for scenarios in which 100 percent, 85 percent, 50 percent, and 0 percent of the fuel supply contained MTBE. The runs were all conducted at the high boating density of 182 boats per day. Predicted peak concentrations were directly proportional to the fuel composition. When 100 percent of the fuel contained MTBE, the predicted peak MTBE concentration in the lake was 6.6 ppb. When MTBE content was reduced to 85 percent and 50 percent, peak seasonal concentrations fell to 5.6 and 3.3 ppb, respectively. Plots of MTBE concentration vs. time for these scenarios are presented Figure 6.9. Restricting the use of fuel containing MTBE, where alternative fuel is available, is seen to be highly effective in reducing the concentrations of MTBE in surface waters

Treatment

MTBE treatment presents specific challenges, including the need to implement a readily available removal technology in a timely manner. Some of the most promising and widely accepted technologies for MTBE treatment include air stripping, advanced oxidation, carbon adsorption, and synthetic resin sorbents. These technologies are proven effective in removing organic compounds. Membranes, biological treatment, and other alternative technologies may be considered at a later date.

Treatment costs depend on existing system conditions. Specifically, treatment costs primarily depend on system flow rates, MTBE influent concentrations, and the level of treatment needed for MTBE to reach the desired effluent goal. Such treatment costs, however, are quite high when compared to the cost of the previously described management options. Treatment costs are further evaluated later in this chapter.

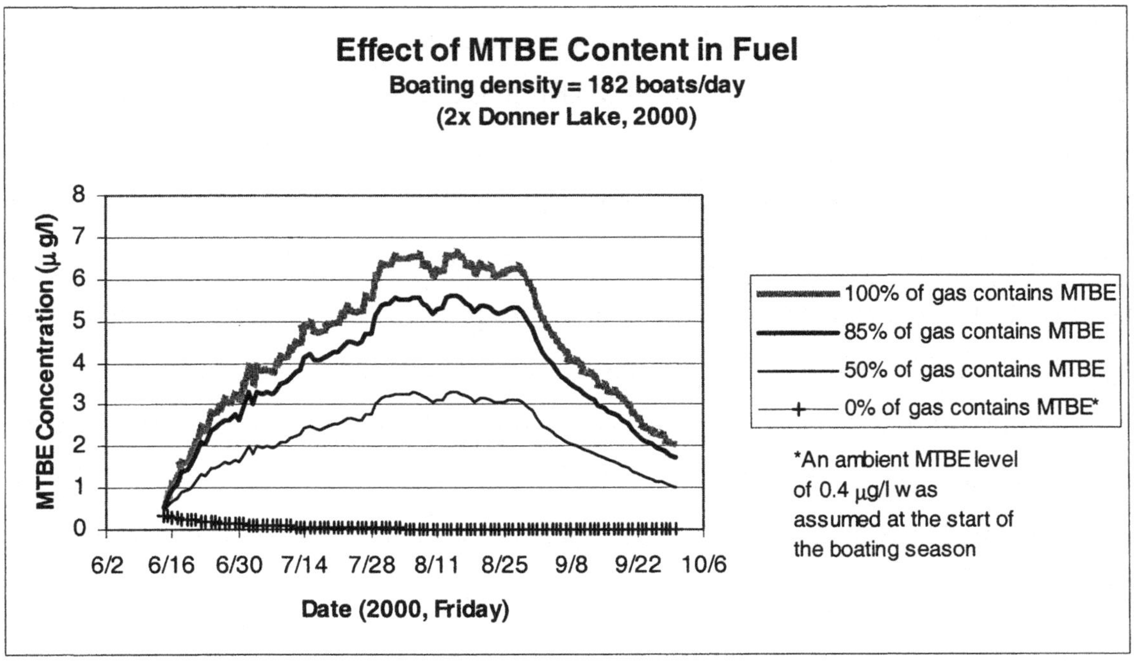

Figure 6.9 Effect of MTBE content in fuel

Applicable Technologies

Because of MTBE's particular physio-chemical properties (such as high solubility, low soil-to-water partition coefficient, and low Henry's Law constant), it is somewhat difficult to remove from water. The treatment of MTBE in drinking water using conventional treatment processes (i.e., air stripping and granular activated carbon [GAC]) poses challenges relative to other organic contaminants — particularly BTEX compounds — because of MTBE's unique physical and chemical properties. The Henry's Law constant of MTBE is approximately an order of magnitude lower than those of BTEX compounds. Since the air stripping potential of a compound is primarily determined by its Henry's Law constant, air stripping of MTBE will tend to be more difficult and more costly relative to BTEX compounds. Furthermore, the relatively low organic carbon-partitioning coefficient and the high water solubility of MTBE make GAC adsorption less effective for MTBE relative to BTEX compounds.

Description of Technologies

The treatment technologies described below are most commonly used to treat groundwater, in which the concentrations of contaminants are higher than are expected in surface water. However, all of these technologies can also be applied to lower concentrations of contaminants. In general, this may be more expensive because the per-unit cost of removing lower concentrations of contaminants is higher than the cost of removing higher concentrations.

Air Stripping. The tendency for a compound to be removed from water by air stripping is characterized by its Henry's Law constant. The Henry's Law constant of MTBE is several times lower than those of other organic compounds commonly treated through air stripping (e.g., trichloroethylene [TCE] and benzene) and, thus, air stripping of MTBE is more difficult and more costly than for these other compounds. However, air stripping is a proven technology that has been used successfully to remove MTBE from drinking water. Packed tower air strippers are being used successfully for drinking water treatment in La Crosse, Kansas and Rockaway Township, New Jersey.

Packed tower aeration was found to be superior to the other air stripping technologies from a cost perspective, regardless of hydraulic capacity, removal efficiency requirements, or initial MTBE concentrations. At higher flow rates (>600 gallons per minute [gpm]) and higher removal efficiencies (>95 percent), packed towers are not only less expensive but often the only technology capable of achieving the treatment goal. However, for lower flow rates (<100 gpm), low profile air strippers become cost competitive with packed towers ($1.80 to $1.86 vs. $1.75 per 1,000 gallons treated, respectively, for 97.5 percent MTBE removal). In addition, low profile air strippers are generally easier to install, maintain, and modify for changing flow and water quality conditions than packed towers. Thus, for hydraulic capacities less than 100 gpm, which may be a remediation or a small drinking water application, a low profile air stripper is generally recommended. For drinking water applications requiring hydraulic capacities greater than 100 gpm, the packed tower aeration technology is recommended. The evaluation of air stripping assumed that off-gas treatment is required when MTBE gas phase concentrations exceed 1 pound per day (lb/day). If off-gas treatment is required and MTBE influent concentrations are low (<200 μg/L), vapor phase GAC is generally the most cost-effective off-gas technology because carbon usage rates are low (as a result of the very dilute MTBE stream) and, thus, operation and maintenance (O&M) costs remain low. If MTBE influent concentrations are higher (e.g., the 2,000 μg/L scenario), oxidation is the recommended technology for an air stream from a packed tower system. The cost analysis indicates a small difference in costs between catalytic and thermal oxidation. Thermal oxidation is the recommended technology to evaluate in conjunction with the selected aeration technology because, like GAC, it is a commonly used and proven technology. Both GAC and thermal oxidation demonstrate equally high levels of reliability, flexibility, removal efficiencies, and cost-effectiveness. Appendix K lists the costs associated with the most cost-effective off-gas treatment technologies, vapor phase GAC, and thermal oxidation. For the assumptions used in the cost analysis of air stripping and off-gas treatment systems, as well as more detailed cost breakdowns, see Chapter 2 of the report (European Fuel Oxygenates Association 2000).

Adsorption. Carbon adsorption technology has been widely used in the past for removal of organic contaminants from drinking water. Simplicity and stable operations are the primary advantages of GAC relative to other water treatment technologies. GAC systems are easy to implement due to wide commercial availability for other applications, and numerous vendors can supply GAC and the necessary equipment (e.g. vessels, piping, and pumps). Because of the simplicity of the equipment and materials, capital and installation costs are relatively low compared to more innovative technologies. Finally, there is no off-gas treatment required for GAC systems and the creation of by-products is limited to spent carbon that can either be thermally regenerated or discarded. However, in spite of these advantages, GAC field-application knowledge specifically for MTBE removal from large-scale drinking water systems is currently limited. The effectiveness of GAC for the treatment of MTBE has been limited by its poor physical and chemical adsorption characteristics. In particular, MTBE's high solubility causes the compound to preferentially remain in solution rather than be adsorbed onto a solid surface.

In addition, natural organic matter (NOM) and other synthetic organic chemicals (SOCs) compete with MTBE for the adsorption sites of GAC. Since MTBE is only weakly adsorbed by GAC, other more preferentially adsorbed SOCs in the contaminated water can result in the desorption or the displacement of previously sorbed MTBE from the GAC matrix. Based on the MTBE isotherms currently available, coconut shell GAC appears to have better MTBE adsorption characteristics than coal-based GAC.

Carbon usage rates and unit treatment costs are highly dependent on influent MTBE concentrations, background water quality, and the concentration of other SOCs. The cost analysis suggests that GAC is most cost-effective for the removal of lower MTBE concentrations, which result in lower carbon usage rates and, consequently, lower O&M costs. GAC is also more likely to be cost-effective for waters that are relatively clean with respect to NOM (e.g., some groundwater). For example, computer modeling predicts that carbon fouling from NOM can cause approximately 50 percent increases in carbon usage rates for the removal of 20 μg/L MTBE. Finally, GAC is more cost-effective for waters contaminated solely with MTBE since other SOCs will preferentially occupy adsorption sites and thereby increase carbon usage rates. Adsorption modeling and cost estimates show that moderate loads of total BTEX (800 μg/L) can cause greater than 50 increases in carbon usage rates for MTBE removal for systems treating influent with 20 μg/L MTBE.

The total amortized costs ($/1,000 gallons; in 1999 $) associated with GAC systems are summarized in Table ES-4. The cost estimates developed for GAC systems are all based on non-detect (<0.5 μg/L) effluent concentrations, resulting in removal efficiencies ranging from 97.5 to >99 percent, depending on influent concentrations. These costs were developed assuming a 30-year design life. It is important to note that because of GAC's high O&M to capital ratio, this technology is more likely to be cost effective for shorter-duration projects.

Advanced Oxidation. Advanced Oxidation Processes (AOPs) destroy MTBE and other organic contaminants directly in the water through chemical oxidation, as opposed to simply transferring them from the liquid phase into a gas phase (in the case of air stripping) or into a solid phase (in the case of GAC and resins). Removal of organic compounds from water by AOPs is primarily accomplished through the reaction of organic contaminants with highly reactive hydroxyl radicals (•OH) that can be produced through a variety of mechanisms. Compared to more established drinking water treatment alternatives, such as air stripping and GAC, AOPs are generally considered an emerging technology. Currently, there are several full-scale applications where organic contaminants (e.g., perchloroethylene [also known as tetrachloroethylene, or PCE] and N-nitroso dimethyl amine [NDMA]) are being removed from drinking water using an AOP. There are, however, no full-scale installations of AOPs for removal of MTBE from drinking water. Thus, thorough pilot- and field-scale testing of an AOP system is required before it can be implemented for MTBE removal in drinking water applications.

Some of the challenges with respect to the implementation of AOPs in drinking water treatment are associated with the formation and fate of oxidation by-products (e.g., tertiary-butyl alcohol [TBA] and tertiary-butyl formate [TBF]), non-selective radical oxidation, radical scavenging, and bromate formation (for ozone-based AOPs). Although it is possible to overcome these challenges, costs will increase as a result of the required greater energy usage, greater chemical dosage, and/or secondary treatment polishing steps. The report (European Fuel Oxygenates Association 2000) evaluated the following established and emerging AOPs potentially applicable for the removal of MTBE from drinking water:

The two most promising AOP technologies appear to be hydrogen peroxide/ozone and hydrogen peroxide/ultraviolet. Both of these processes are well understood and have been demonstrated at several bench- and field-scale sites to successfully remove MTBE from water to meet drinking water standards.

In addition to these two relatively well-established AOPs, E-beam and cavitation are two emerging AOPs that warrant further consideration due to their technical feasibility for removing MTBE from drinking water to meet standards. These technologies are still in their commercial developmental stages for removal of organic contaminants in drinking water applications; however, they have been widely demonstrated for disinfection and remediation applications.

Besides being the most technically feasible, hydrogen peroxide/ozone and hydrogen peroxide/ultraviolet (in addition to cavitation) appear to be the most economically feasible. However, these costs are strongly dependent on source water quality and are difficult to verify due to the untested nature of these technologies in large-scale applications. Cavitation costs involve the most uncertainty because there are no pilot, field, or full-scale drinking water treatment applications for MTBE removal. Consequently, while there is significant uncertainty for all the cost estimates, H2 O2 /O3 and H2 O2 /MP-UV technologies are essentially equivalent in cost and less expensive than the other AOPs evaluated.

Emerging technologies include:

- High Energy Electron Beam Irradiation (E-beam)
- TIO_2 -catalyzed UV

- Sonication/Hydrodynamic Cavitation
- Fenton's Reaction

Biological Treatment. Biological degradation of organic contaminants is common in wastewater treatment. Biological MTBE removal converts MTBE to cell mass, carbon dioxide, and water by microbial cultures. MTBE in the liquid or air phase is brought into contact with microbial cultures that are either attached to a support media or suspended in liquid and use MTBE as a carbon and energy source. MTBE from the contaminated liquid or air diffuses into the thin film and is degraded. The liquid film in biofilters is maintained by adding water to keep the bed moist. For the aerobic process dissolved oxygen is added in the influent or diffused through the thin film. Nutrients are added if the contaminated flow does not contain adequate amounts to support biological activity. The removal rate of MTBE depends strongly on the composition of the water to be treated. In addition, there is concern over the reliability and microbial quality of treated water.

Membrane Separation. Membrane processes require a membrane is able to reject or select passage of certain dissolved species based on compound size, shape, and/or charge. Membrane processes include reverse osmosis, electrodialysis, and ultrafiltration. Membrane processes are generally considered in cases such as desalination, brackish water conversion, and removal of specific ions that are difficult to remove with other processes. Removal efficiency of membrane process depends on the quality of the feed water and desired product water quality; the more contaminated the feed water and the higher the desired product water quality, the greater the likelihood of membrane fouling (Hamman et al. 1990). Though reverse osmosis and nanofiltration are possible processes to selectively remove MTBE from water, reverse osmosis has the greatest potential for MTBE removal because it is designed to remove small solutes such as sodium and chloride ions. Since no information is available on MTBE removal efficiency with membrane separation, it will not be considered further in the analysis.

Despite the relatively proven nature of the treatment processes discussed above, many treatment and remediation scenarios may require a combination of treatment processes to meet effluent criteria. For example, all drinking water treatment processes in California will require redundant treatment (i.e., a treatment process that serves as a barrier between the consumer and the contamination). In many cases, GAC will be used as the redundant process. Similarly, AOPs will often employ additional treatment processes to remove oxidation by-products or residual oxidants. Alternatively, some singular treatment processes may not be able to reach drinking water effluent criteria, especially in cases where the influent concentrations are high (e.g., 2,000 μg/L) and the effluent criteria is non-detect. In these circumstances, two treatment processes may be required — the first process would be used to lower concentrations by 90 to 99 percent and the second process would be used as a polishing step to remove the remaining 1 to 10 percent.

Operations Management

MTBE reduction in the lake and reservoir is also possible through operational considerations. This could involve changing the location of the intake / water supply withdrawal or modifying the timing of withdrawal. Both of these are a function of seasonal variations and water agency capabilities. The relation of the intake/water supply structures to the water quality distribution in the reservoir also varies with local climate conditions. In times of low water level, for example after a prolonged drought, previously deep, hypolimnetic intake structures may actually encounter epilimnetic water. For these reasons it is difficult to make generic recommendations for operations management based on model runs for one particular reservoir.

To demonstrate the effect of location of withdrawal structures, estimates were made of the effect of lake overturn on MTBE concentration. The DLM-MTBE model was used to estimate the concentration of MTBE required in the epilimnion that would create a problem due to lake overturn. This will be useful in locating the intake point. It should be noted that these results are specific to Donner Lake, and variations in the hypsographic relationships at other reservoirs and lakes will produce different results.

Two concentrations were considered: (i) 0.5 ppb (MTBE detection limit) and (ii) 5 ppb (California secondary MCL). Model predictions indicate that for a 0.5 ppb concentration, the required concentration in the epilimnion before turnover is 1.09 ppb; for a 5 ppb concentration, the value is 10.86 ppb. Therefore, if two intake / withdrawal locations are to be considered based on seasonal fluctuations, the MTBE concentrations in the epilimnion would need to be maintained below these levels.

Location of Water Supply Withdrawals

Many water agencies control their source operations through use of a multi-level outlet tower. In many cases, water can be selected at various levels which can be used to help control levels of MTBE in the water sent to treatment plants or other downstream use. Generally, utilities select water from the hypolimnion since the epilimnion can be more difficult to treat. The epilimnion is prone to algal blooms, which tend to clog filtering operations and produce related taste and odors. Fortunately, for those who have this capability it is possible to profile the reservoir and select zones with the lowest concentrations of MTBE.

One serious problem with using hypolimnion water is that reservoirs can become anaerobic. This causes higher iron and manganese levels, as well as producing hydrogen sulfide that imparts a rotten egg smell and an oxygen deficit to the water column. Some reservoir profiles are deep enough to allow the lower reach to remain anoxic while still providing water from a zone that continues to support an oxygen level. Aeration and hypolimnetic oxygenation may be considered as a tool to prevent the anoxic conditions that can be used in conjunction with selecting water levels for MTBE control.

Reservoirs that are shallow and do not support a sufficient volume of hypolimnetic water to meet water supply needs are obviously hampered in terms of using this approach. In these instances, it is not possible to use the depth profile of the reservoir to control MTBE.

Timing of Water Supply Withdrawals

Modifying the time of year for source water withdrawals is very dependent on the specific situation and capability of water utilities. If it is possible to limit source water withdrawal during the off peak boating season, then this method may help in MTBE control. Boating density and timing of boat use may also be considered to create windows of operation to avoid peak MTBE presence. However, most water utilities take water during peak demands or year round, which overlap with recreational boating. Some water agencies are bound by water contracts which may be favorable to water use from the reservoir during off peak boating season and may have multiple sources which can be used in an overall approach to control MTBE.

Timing of water supply withdrawals is just one tool that may enable a water utility to develop a strategy for dealing with MTBE or other fuel related oxygenates.

Use of operational strategies for MTBE control should not be underestimated since timing of water withdrawal, timing of boating activities, and ability to take water at various levels in a reservoir have the capability to control MTBE to a tolerable level. In many cases this may be preferable to instituting a ban on recreational boating. Each of these concepts has been briefly discussed here since each water agency has specific circumstances and limitations. The key to using operational strategies is to develop a historical database of time-based occurrence in order to identify opportunities to make use of various ideas. Operational strategies are also subject to seasonal variations and some best judgement or conservative factors may be necessary to ensure regulatory compliance, which is dependent on the degree of variation occurring in MTBE presence. Presence of MTBE should be monitored on a regular basis in both the epilimnion and hypolimnion during database collection to ensure that enough information is available to develop operational strategies.

Economic Considerations

It is clear from the preceding material that owners/operators of lakes and reservoirs have a number of options for reducing levels of MTBE and other fuel related contaminants in their water bodies. Prior to making a decision as to the option most appropriate for them, it is necessary to weigh the cost of implementing the option against the loss or potential loss of economic activity that may result. The purpose of this section is to document how overall economic considerations can be evaluated in such cases, using the well-documented Donner Lake case as an example.

While the economic value of boating activity for a given lake or reservoir can vary extensively, it is believed that Donner Lake is a very typical case. The size, location, and level of boating activity at Donner are well documented elsewhere in this report, allowing operators of different water bodies a reasonable basis of comparison.

Recreational boating contributes significant economic activity in those regions where it is enjoyed. It is not difficult to conduct a regional economic impact analysis of recreational boating. Appendix J contains an evaluation of the regional economic impacts of recreational boating in

Donner Lake. Table J.2 (Appendix J) shows the total employment resulting from recreational boating in Donner Lake to be 188 jobs, and the total annual income to be $3.4 million. This economic impact represents about 0.7 percent of the total regional industry employment in Nevada County, CA. Thus recreational boating has a significant beneficial impact on regional income and personal employment in the county.

Since many of the source control options for managing water quality involve eliminating or reducing boating activity, it is necessary to at least compare these adverse economic impacts when making a decision as to which option to implement. Both direct costs to the agency or water utility and economic impacts to the region need to be considered.

Direct costs will primarily be to the water utilities, which will include monitoring, enforcement, and compliance costs. Items contributing to this cost would typically be boats, enforcement personnel, gates, and security. In cases where water treatment or intake management are adopted, the direct costs are the amortized capital and annual operating costs of the facilities that will need to be constructed and operated in order to implement those options.

Economic impacts are to the region due to loss of economic activity. Such impacts could result in total loss of the economic value of recreational boating, as in the case of a total ban on all boats, or they could result in only a partial loss of such economic value as in the case of boating restrictions. In the case of a ban on unmodified two-stroke engines, the economic loss could be a temporary one during which time boat owners slowly replace their conventional two-stroke engines for engines with advanced technology. For those replacement engines purchased in Nevada County, an economic benefit would actually accrue.

According to 1998 figures, replacement engines can cost between $4,000 to $15,000 depending upon the size (California Bowen Bill 1998). The East Bay Municipal Utility District (EBMUD) placed a restriction on polluting two-stroke engines on the San Pablo Reservoir beginning January 2000. To ensure the effectiveness of the restriction, EBMUD spent approximately $400,000 in August-September 1999 to replace EBMUD-owned rental boat outboard engines (Berger 2001).

Economic Impact of MTBE Management Options

This section presents the costs and economic impacts associated with several of the management options, using Donner Lake, California as the example case.

Costs

The costs for the various source control and treatment options are shown in Table 6-1. Monitoring costs are from the Keller et al. (1998) report and a study done by Tikkanen (1998) at EBMUD reservoirs that estimated monitoring costs for most surface water reservoirs at $10,000 to $25,000 per year per reservoir. These costs are based on the number of samples taken per month (typically 10 per reservoir), analytical cost (ranging from $50 to $100 per sample), and the associated labor and boating expenses incurred during the sample collection process (between $500 to $800 per month). For those cases involving restrictions or partial banning of boating activity, an estimated $25,000-50,000 per year is added for patrolling and enforcement.

Treatment costs are adapted from Appendix K and are based on the unit costs associated with a 6,000 gpm flow, 20 ppb MTBE influent concentration, and 0.5 ppb effluent concentration. The influent concentration chosen was 20 ppb because this is the taste and odor threshold for MTBE. The same treatment technologies can be used at lower concentrations, but the per-unit removal cost may be higher than it would be for higher concentrations. For Donner Lake, it is assumed that 6,000 gpm or 8.6 million gallons per day is an appropriate maximum average daily withdrawal for drinking water purposes.

Costs of raw water intake modification and operation can be quite variable depending upon the characteristics of a lake or reservoir and its existing intake. Generally, the amortized capital costs plus operating cost of intake modification will be less than a treatment option.

Using Donner Lake as an example, capital costs for intake are estimated to range from $200,000 to $300,000 while annual O&M costs are estimated to range from $4,000 to $6,000 (Kaufman 2001). These cost estimates are based on the assumption that the length and depth of the intake are approximately 180 feet and 50 feet respectively.

Economic Impacts

The economic impacts shown in Table 6.1 are reductions in the base economic value of recreational boating in Donner Lake as presented in Appendix J. The two boating restriction scenarios shown are a ban on all conventional (non-direct injection) two-stroke engines and a 35 percent reduction in boating activity, regardless of type of engine. For Donner Lake, since conventional two-stroke engines comprise approximately 35 percent of all engines using the lake, the immediate economic impact for both scenarios is the same. In the case of the ban on conventional two-stroke engines, it is assumed that economic value is recovered over time as old engines are replaced with new ones.

Table 6.1

Cost considerations of options to manage MTBE in lakes and reservoirs

Management option	Direct Cost	Loss of economic value
Ban two-stroke[1]	$ 10,000-25,000/year monitoring $25,000-50,000/year enforcement	Short-term impact: 66 jobs and $1.1 million total income Long-term impact: Negligible to slightly positive depending on how quickly engines are replaced.
Boating density reduced by 35 percent[2]	$ 10,000-25,000/year monitoring $25,000-50,000/year enforcement	Short-term impact: $0.3 to $0.8 Million 15 to 51 jobs Long-term impact: Negligible to slightly positive depending on how quickly engines are replaced.

(continued)

Table 6.1 (continued)

Management option	Direct Cost	Loss of economic value
Treatment		
Assume 6,000 gpm treated flow, 20 ppb MTBE influent, 0.5 ppb effluent	$900,000-$3.3 million/year depending on technology used	None
Unit costs from Appendix K		
Intake Modification	$20,000-30,000/year (amortized cost, assuming $200,000-$300,000 capital cost, O&M cost of $4000-$6000/year, and 7 percent interest over 30 years)	None
	$10,000 to $25,000 per year per reservoir monitoring	

[1] Using Donner Lake example

[2] Using Donner Lake example where two-stroke engines represent 35 percent of all engines (per 2000 boating survey)

[3, 4, 5] See Appendix K

APPENDICES

APPENDIX A
LIST OF VOCS

Table A.1

VOCs emitted by recreational watercraft

Compound	Retention Time (minutes)	Maximum Contaminant Level Goal (MCLG) *mg/L*	Maximum Contaminant Level (MCL) *mg/L*
Benzene	7.48	Zero	0.002
Toluene	11.38	1	1
Ethylbenzene	14.96	0.7	0.7
p-Xylene	15.32	10 (total Xylenes)	10 (total Xylenes)
m-Xylene	15.32	10 (total Xylenes)	10 (total Xylenes)
Phenylacetylene	15.96	--	--
o-Xylene	16.20	10 (total Xylenes)	10 (total Xylenes)
Styrene	16.27	0.1	0.1
Cumene	17.26	--	--
Proplylbenzene	18.31	--	--
4-Ethyltoluene	18.64	--	--
3-Ethyltoluene	18.64	--	--
Mesitylene	18.88	--	--
two-Ethyltoluene	19.25	--	--
Benzaldehyde	19.57	--	--

(continued)

Table A.1 (continued)

Compound	Retention Time (minutes)	Maximum Contaminant Level Goal (MCLG) *mg/L*	Maximum Contaminant Level (MCL) *mg/L*
Pseudocumene	19.81	--	--
m-Methylstyrene	20.00	--	--
p-Methylstyrene	20.11	--	--
Benzofurane	20.23	--	--
1,4-Dichlorobenzene (I.S.)	20.50	--	--
1-Isopropyl-3-methylbenzene	20.70	--	--
Hemellitene	20.80	--	--
Indane	21.15	--	--
1,3-Diethylbenzene	21.62	--	--
Indene	21.70	--	--
1-Methyl-4-proplylbenzene	21.79	--	--
1-Methyl-3-proplylbenzene	21.79	--	--
1,4-Diethylbenzene	21.81	--	--
Butylbenzene	21.85	--	--
1,3-Dimethyl-5-ethylbenzene	21.94	--	--
1,2-Diethylbenzene	22.03	--	--
1-Methyl-2-propylbenzene	22.24	--	--

(continued)

Table A.1 (continued)

Compound	Retention Time (minutes)	Maximum Contaminant Level Goal (MCLG) *mg/L*	Maximum Contaminant Level (MCL) *mg/L*
Salicylaldehyde	22.32	--	--
1,4-Dimethyl-2-ethylbenzene	22.58	--	--
1,3-Dimethyl-4-ethylbenzene	22.62	--	--
1,2-Dimethyl-4-ethylbenzene	22.85	--	--
o-Tolualdehyde	23.05	--	--
m-Tolualdehyde	23.23	--	--
Phenol	23.29	--	--
Methylindane	23.38	--	--
Methylindane	23.49	--	--
1,2-Dimethyl-3-ethylbenzene	23.55	--	--
p-Tolualdehyde	23.70	--	--
Durene	23.88	--	--
Isodurene	24.00	--	--
Methylindane	24.35	--	--
Methylindane	24.55	--	--
Methylindane	24.96	--	--
1,2,3,4-Tetramethlbenzene	25.08	--	--
Methylindene	25.14	--	--

(continued)

Table A.1 (continued)

Compound	Retention Time (minutes)	Maximum Contaminant Level Goal (MCLG) *mg/L*	Maximum Contaminant Level (MCL) *mg/L*
Methylindene	25.40	--	--
plm-Cresol	26.00	--	--
Dimethylindane	26.12	--	--
Dimethylindane	26.23	--	--
Napthalene[1]	26.50	--	--
2,5-Dimethylbenzaldehyde	26.61	--	--
Dimethylbenzaldehyde	26.73	--	--
Dimethylbenzaldehyde	26.85	--	--
C_5-benzene	26.87	--	--
2,4-Dimethylbenzaldehyde	26.96	--	--
Methylbenzofurane	27.07	--	--
3,4-Dimethylbenzaldehyde	28.09	--	--
2-Nethylnapthalene	29.85	--	--
1-Monthylnapthalene	30.32	--	--

Data from Juttner 1995

Notes:
[1] Napthalene is included in the list of "Regulatory Determination Priorities." That is, a decision as to whether to regulate Napthalene will be made by USEPA by August 2001.

APPENDIX B
DETAILED DESCRIPTION OF MODEL

HYDRODYNAMIC MODEL

The DLM was developed in 1978 as a one-dimensional simulation model of the vertical distribution of temperature and salinity in small- to medium-size lakes and reservoirs. The assumption of one dimensionality means that variations in density, temperature, and VOC concentration in the lateral directions are assumed to be small compared with variations in the vertical. The core of DLM is a process-based hydrodynamic model which parameterizes the individual transport and mixing processes in a lake and sequentially couples them to predict the lake's density structure over time (McCord and Schladow, 1998). The following discussion applies to both the 1998 and 2002 versions of DLM.

The model, as configured for the current study is based on an earlier series of reservoir models (Imberger et al. 1978; Imberger and Patterson 1981; Hamilton and Schladow 1996; Schladow and Hamilton 1996).

The vertical profile of the lake is represented as a set of up to 200 Lagrangian layers which are free to move vertically, and to contract and expand in response to inflows, outflows and surface mass fluxes. The Lagrangian formulation avoids the need to calculate vertical velocities, greatly decreasing computational time and minimizing numerical diffusion. Each layer is homogeneous, and property differences between layers represent the vertical distribution. Layer thicknesses are adjusted within the model according to the resolution required to represent the vertical density gradient. The model is initialized with measured data for each layer.

Mixing is represented by the amalgamation of layers. Properties of the amalgamated layer are volumetrically averaged, and the total number of model layers is decreased accordingly. Amalgamated layers may be split according to a specified maximum layer thickness criterion. Conversely, when a layer size falls below a specified minimum criterion, as may occur when there is withdrawal of water from the lake, then the layer is amalgamated with the smaller of the two bounding layers. Inflows are inserted at their level of neutral buoyancy, after allowance for entrainment has been made. To model withdrawal of water from a lake, a withdrawal layer thickness and location is calculated based on the height and type of outlet port, the local density stratification at that height, and the rate of withdrawal.

Daily meteorological conditions, stream inflow, stream outflow (or reservoir withdrawals), and initial water quality conditions are used as input to the model. The model outputs daily vertical profiles of temperature and MTBE concentration in the lake, and temperature and MTBE concentration of the outflowing (or withdrawn) water.

Whether a one-dimensional representation is appropriate for a particular lake may be estimated by evaluating four dimensionless parameters (Patterson et al. 1984). The force balances that comprise these parameters relate to the surface mixing processes, rotational effects, stream inflows, and submerged discharges. They may be defined as:

$$W = \frac{g' h_S}{u_*^2} \frac{h_S}{L} \tag{1}$$

$$R = \frac{(g' h_s)^{1/2}}{f B} \tag{2}$$

$$F_I = \frac{V}{(g' h_s)^{1/2}} \tag{3}$$

$$F_O = \frac{Q}{H^2 (g' h_s)^{1/2}} \tag{4}$$

In Equation (1) for the Wedderburn number, L is the length of the reservoir, h_S is a representative surface mixed layer depth, g', the effective gravitational acceleration, determined by replacing the continuous field density stratification with an equivalent two-layer stratification, is the density step across the two layer system, and u_* is the surface shear velocity. For one-dimensional surface mixing processes to dominate, the appropriate value of the criterion is $W > 1$.

In Equation (2) for the Rossby number, f is the Coriolis frequency, B is the lake width and g' and h_S are as defined above. $R > 1$ is the criterion for the absence of rotationally induced two-dimensional effects.

The internal Froude number defined in Equation (3) represents the balance between the inertial forces associated with an inflow, and the pressure forces induced by horizontal density gradients generated by the inflow. V is the inflow velocity, and g' is based on the difference between the inflow and the surface densities. For $F_I < 1$, the pressure forces adjust the gradients faster than the inflows can generate them, and the structure remains one-dimensional.

The Froude number based on outflow, F_O, also reflects the relative importance of inertial and gravitationally induced pressure forces. In Equation (4), Q is the withdrawal flow, H is the total reservoir depth, and g' is based on the density difference between the two layers. For $F_O < 1$, the one-dimensional structure remains intact.

The one-dimensional model has been shown to be appropriate for all the lakes simulated in this study (McCord and Schladow 1998).

VOC VOLATILIZATION SUB-MODEL

Volatilization of volatile organic compounds from the surface of a lake is modeled as a two-film diffusion process, involving resistance in both the air and water phases. Diffusion rates in both the surface layer of water and the overlying air mass are important, although the overall rate for fuel-based contaminants tends to be limited by the liquid-side transfer velocity. The overall air-water transfer velocity, k_{OL} (m s^{-1}), is given by Lewis and Whitman (1924):

$$\frac{1}{k_{OL}} = \frac{1}{k_L} + \frac{RT_w}{H\,k_G} \qquad (5)$$

where R is the universal gas constant (8.2×10^{-5} atm m^3 mol^{-1}), and H is Henry's Gas Law coefficient (atm m^3 mol^{-1}). The formulation for calculating H is that of Robbins et al. (1993),

$$H = e^{(a - b/T_w)} \qquad (6)$$

where T_w is the surface water temperature (°K). The coefficients a and b for various compounds are given by Robbins et al. (1993). For MTBE, this yields slightly different values than those reported by USEPA (1986). For example, at 20°C H given by Eqn (6) is $4.3\text{x}10^{-4}$ atm m^3 mol^{-1}, compared to the USEPA value of $4.5\text{x}10^{-4}$ atm m^3 mol^{-1}.

Wind speed is functionally related to both the gas-side and liquid-side transfer velocities. The transfer velocity (cm s^{-1}) in air, k_G, is calculated following Schwarzenbach (1993), as given by Asher (2002):

$$k_G = \left(\frac{D_a(VOC)}{D_a(H_2O)}\right)^{\frac{2}{3}}(0.15U_w) \tag{7}$$

where U_w is wind speed in m s^{-1} and $D_a(VOC)$ and $D_a(H_2O)$ are the diffusivities in air of the compound and water respectively. The ratio of the diffusivities is given as (Atkins 1986 as presented in Asher 2002):

$$\frac{D_a(VOC)}{D_a(H_2O)} = \sqrt{\frac{18}{M_{VOC}}}\left(\frac{T_a}{293.16}\right)^{3/2} \tag{8}$$

where M_{VOC} is the molecular weight of the compound of interest, and T_a is air temperature.

The transfer velocity (cm s^{-1}) in water, k_L, is calculated following Wanninkhof et al. (1991), in Asher (2002):

$$k_L = 0.45\sqrt{\frac{600D_L}{\nu}}\,U_w^{1.64} \tag{9}$$

where ν is the kinematic viscosity of water, in cm^2 s^{-1}., and D_L is the molecular diffusivity of the compound in water, also in cm^2 s^{-1}. D_L is calculated according to Hayduk and Laudie (1974) in Asher (2002):

$$D_L = \frac{7.48\times10^{-8}\sqrt{2.26M_{H2O}}\ T_w}{\mu V^{0.6}} \tag{10}$$

M_{H2O} is the molecular weight of water, T_w is water temperature in Kelvins, μ is the absolute viscosity of water in centipoise (Weast 1983), and V is the molar volume of the compound. V is calculated by dividing the molar weight by the density, which for MTBE yields:

$$\frac{88.15\ \text{g/mol}}{0.7405\ \text{g/cm}^3} = 119.0\ \text{cm}^3\text{/mol}$$

Finally, the change in concentration of the compound (*dC/dt*) in the surface layer owing to volatilization is computed as

$$\frac{dC}{dt} = \frac{k_{OL}\,(Cs - Ct)}{h} \tag{11}$$

where C_t is the concentration in the surface layer and C_s is the saturation concentration of the compound in water, at equilibrium with an assumed atmospheric concentration.

It should be noted that the variable *h* in Eqn (11) is the depth of the uppermost computational layer, typically less than 1 m thick in model simulations. At times of active mixing (e.g. at night or during high wind conditions), this uppermost layer is amalgamated with underlying layers at each 15 minute time step. Thus the concentration of MTBE in the epilimnion would be observed to decrease throughout its depth (in the absence of new inputs). By contrast, on warm days when the epilimnion assumes a diurnal stratification, there would be a depletion of just the near surface concentration of MTBE. This is consistent with observations using sulfur hexafluoride gas tracers. It is to be distinguished from the limnological surface layer of the lake, which may have a depth of tens of meters, and is represented in the model as a combination of a number of computational layers.

Transport of MTBE between computational layers is either by advection, if there is sufficient turbulent kinetic energy, or by diffusion for stably stratified regions of the water column. The diffusion coefficient used is the summation of the molecular diffusivity of MTBE, described above, plus the turbulent diffusivity. The value of the turbulent diffusivity coefficient depends directly on the dissipation of the turbulent kinetic energy and inversely on the stratification (Weinstock 1981). In general, the diffusion is dominated by the turbulent component of the diffusivity, which is typically several orders of magnitude greater than molecular values.

Degradation sub-model

In the process of calibrating DLM-2002, it became apparent that volatilization alone could not account for the measured levels of MTBE and toluene in Donner Lake. Modeled levels of these two compounds were consistently higher than measured levels. MTBE, while problematic, was less affected than toluene because of its greater resistance to degradation, but it was clear that some other loss mechanism was operating, most likely microbial degradation.

An initial assumption was made that the most important process was aerobic microbial degradation, and that it followed first order kinetics. Degradation rates were obtained from Rathbun (1998), which followed the work of Howard et al. (1991). There is limited research in this area, and most of the published rates are estimates, based on what experimental data are available. Because of the uncertainty, the rates are given as a range. For xylene, for example, the half-life ranges between 7 and 28 days. Rates for benzene, toluene, and ethylbenzene are given as first-order rate coefficients, and the rate for benzene is based on experimental data. For MTBE and the xylenes, rates are given as half-lives. Half-lives were converted to pseudo-first-order coefficients as follows (Rathbun 1998):

$$k_t^{'} = \frac{0.693}{t_{0.5}} \quad (t^{-1})$$

where $t_{0.5}$ is the half-life in the desired time units.

A first order coefficient of this sort necessarily amalgamates a great many individual processes into a single variable. The nature of the individual processes cannot be known, and the bulk action is inferred. Some may be aerobic, some anaerobic. Nevertheless, the formulation was able to reasonably represent conditions encountered at the three selected reservoirs.

The high value from the given range for each compound was chosen initially, and this turned out to provide a good fit with the measured data. Given the wide range of values and the associated uncertainty, applying temperature corrections to the rates is of secondary importance. Nevertheless, model results were more consistent when the degradation rates were adjusted for water temperature. A temperature of 20°C was assumed for the published rates, and adjustments were made using the Arrhenius equation, with a value of 1.05 for the correction factor theta. The degradation coefficient is then computed as:

$$k_T = k_{20}\theta^{(T-20)}$$

where

k_{20} = the given degradation coefficient

□= correction factor, = 1.05

T = water temperature

k_T = the adjusted degradation coefficient

The new concentration is computed by:

$$C = C_0 e^{-k_T t}$$

where

C_0 = initial concentration

C = final concentration

t = time (time of model step, in same units used for the degradation coefficient)

VOC loading

VOC influx to the surface layer of a lake has been assumed to be entirely a function of boat use. The VOC load depends on the total number of boats using the lake/reservoir (*N*); the distribution of their engine types; the time of operation, fuel consumption, and emission rates of each type; the types of boating activity; and the volume fraction of each VOC in gasoline. An average VOC load was computed for each of the 4 boating season months (June-September), and for weekend and midweek days. The load for each day modeled was calculated by computing the ratio of that day's boating activity level to the average level, and using the ratio as a multiplier for the average VOC load. The model divides the daily VOC load equally within the hours 6 a.m. – 6 p.m., when boating is most common.

APPENDIX C
LIST OF AGENCIES CONTACTED

Respondents to the MTBE and other fuel-related contaminants questionnaire

Utility	Reservoir Name	City	County
Arrowhead Lake Association	Lake Arrowhead	Lake Arrowhead	San Bernardino, California
Bass Lake Water Company	N. Fork of Willow Creek-Trib. To Bass Lake	unincorporated area	Madera Co., California
Big Bear Municipal Water District	Big Bear Lake	Big Bear Lake	San Bernardino, California
California Water Service Company – Atherton	Bear Gulch	Atherton	San Mateo, California
California Water Service Company – Oroville	Cherokee Reservoir	Oroville	Butte, California
California Water Service Company –Oroville	Oroville Reservoir	Oroville	Butte, California
California Water Service Company – Bakersfield	Lake Isabella	Isabella	Kern County, California
California-American Water Company	Los Padres	Carmel Valley	Monterey Co., California
California-American Water Company	San Clemente	Carmel Valley	Monterey Co., California
Calleguas Municipal Water District	Lake Bard	Thousand Oaks	Ventura Co., California
Casitas Municipal Water District	Lake Casitas	Ventura	Ventura Co., California
City of Folsom, Ashland and Folsom Division	Willow Hill	Folsom	Sacramento, California

(continued)

Respondents to the MTBE and other fuel-related contaminants questionnaire (continued)

Utility	Reservoir Name	City	County
City of Lakeport	Clear Lake	unincorporated area	Lake Co., California
City of San Diego, Water Dept. (Water Quality Lab)	Barrett	unincorporated area	San Diego Co., California
	El Capitan	Lakeside	San Diego Co., California
	Hodges	Escondido	San Diego Co., California
	Lower Otay	Chula Vista	San Diego Co., California
	Miramar	San Diego	San Diego Co. California
	Morena	Campo	San Diego Co. California
	Murray	La Mesa	San Diego Co. California
	San Vicente	Lakeside	San Diego Co. California
	Sutherland	Ramona	San Diego Co. California
City of Santa Barbara PWD, Water Resources Division	Gibralter	City of Santa Barbara	Santa Barbara Co. California

(continued)

Respondents to the MTBE and other fuel-related contaminants questionnaire (continued)

Utility	Reservoir Name	City	County
City of Santa Cruz	Loch Lomond	Felton	Santa Cruz California
City of Vallejo	Lake Curry	unincorporated area	Napa Co. California
	Lake Frey	unincorporated area	Napa Co. California
	Lake Madigan	unincorporated area	Napa Co. California
Contra Costa Water District	Los Vaqueros Reservoir	Byron	Contra Costa Co. California
County of San Luis Obispo	Salinas Reservoir	unincorporated area	San Luis Obispo Co. California
Elsinore Valley Municipal Water District	Railroad Canyon Reservoir (Canyon Lake)	City of Canyon Lake	Riverside Co. California
Humboldt Bay Municipal Water District	Ruth Lake	Ruth	Trinity Co. California
LACC Mutual Water Co.	Lake Almanor	Lake Almanor	Plumas Co. California
Lake Alpine Water Company	Bear Lake	Bear Valley	Alpine Co. California
Lake Don Pedro Community Services Dist.	Lake McClure	unincorporated area	Mariposa Co. California

(continued)

Respondents to the MTBE and other fuel-related contaminants questionnaire (continued)

Utility	Reservoir Name	City	County
Lake Hemet Municipal Water	Lake Hemet	Garner Valley	Riverside Co. California
Lopez Lake - San Luis Obispo Co.	Lopez Lake	Arroyo Grande	San Luis Obispo Co. California
Los Angeles Dept of Water and Power	Long Valley Reservoir (Crowley Lake)	unincorporated area	Mono Co. California
Mammoth Community Water District	Lake Mary	Mammoth Lakes	Mono Co. California
Marin Municipal Water District	Alpine	unincorporated areas	Marin Co., California
Marin Municipal Water District	Bon Tempe	unincorporated area	Marin Co., California
Marin Municipal Water District	Kent	unincorporated area	Marin Co., California
Marin Municipal Water District	Lagunitas	unincorporated area	Marin Co., California
Marin Municipal Water District	Nicasio	unincorporated area	Marin Co., California
Marin Municipal Water District	Phoenix	unincorporated area	Marin Co., California
Marin Municipal Water District	Soulajule	unincorporated area	Marin Co., California

(continued)

Respondents to the MTBE and other fuel related contaminants questionnaire (continued)

Utility	Reservoir Name	City	County
Nacimiento Water Company	Lake Nacimiento	San Louis	Monterey Co., California
Nevada Irrigation District	Bowman	Near Graniteville	Nevada Co., California
Nevada Irrigation District	Combie	Near Auburn	Placer & Nevada Co., California
Nevada Irrigation District	Faucherie	Near Cisco Grove	Nevada Co., California
Nevada Irrigation District	French	Near Cisco Grove	Nevada Co., California
Nevada Irrigation District	Jackson	Near Sierra City	Nevada Co., California
Nevada Irrigation District	Jackson Meadows	Near Sierra City	Sierra & Nevada Co., California
Nevada Irrigation District	Milton	Near Sierra City	Sierra & Nevada Co., California
Nevada Irrigation District	Rollins	Near Colfax	Nevada & Placer Co., California
Nevada Irrigation District	Sawmill	Near Graniteville	Nevada Co., California
Nevada Irrigation District	Scotts Flat	Near Nevada City	Nevada Co., California

(continued)

Respondents to the MTBE and other fuel-related contaminants questionnaire (continued)

Utility	Reservoir Name	City	County
North Coast County Water District	No Reservoir (shallow creek)		
North Marin Water District	Stafford	unincorporated	Marin Co., California
North Tahoe Public Utility District	Lake Tahoe	various	Washoe, El Dorado, Placer, California
Palmdale Water District	Littlerock Dam	Littlerock	Los Angeles Co., California
Palmdale Water District	Palmdale Lake	Palmdale	Los Angeles Co., California
Paradise Irrigation District	Magalia	Magalia	Butte Co., California
Paradise Irrigation District	Paradise Lake	Magalia	Butte Co., California
Redwood Valley County Water District	Redwood Valley County Water District	Redwood Valley	Mendocino, California
San Francisco Public Utilities Commission	Calaveras	Alameda	Santa Clara Co., California
San Francisco Public Utilities Commission	Cherry	unincorporated area	Tuolomne Co., California
San Francisco Public Utilities Commission	Hetch-Hetchy	unincorporated area	Tuolomne Co., California

(continued)

Respondents to the MTBE and other fuel-related contaminants questionnaire (continued)

Utility	Reservoir Name	City	County
San Francisco Public Utilities Commission	Lake Merced (incl. N,S and E Lakes)	unincorporated area	San Francisco Co., California
San Francisco Public Utilities Commission	Lower Crystal Springs	unincorporated area	San Mateo Co., California
San Francisco Public Utilities Commission	Pilarcitos	unincorporated area	San Mateo Co., California
San Francisco Public Utilities Commission	San Andreas	unincorporated area	San Mateo Co., California
San Francisco Public Utilities Commission	San Antonio	unincorporated area	Alameda Co., California
San Francisco Public Utilities Commission	Upper Crystal Springs	unincorporated area	San Mateo Co., California
San Jose Water Company	Lake Elsman	Los Gatos	Santa Clara Co., California
San Juan Water District	Folsom Lake	Granite Bay	Placer and Sacramento, California
Santa Clara Valley Water District	Anderson	unincorporated	Santa Clara County, California
Santa Clara Valley Water District	Calero	unincorporated area	Santa Clara County, California

(continued)

Respondents to the MTBE and other fuel-related contaminants questionnaire (continued)

Utility	Reservoir Name	City	County
Santa Clara Valley Water District	Coyote	unincorporated area	Santa Clara County, California
Stockton East Water District	Goodwin	unincorporated area	Stanislaus Co., California
Stockton East Water District	New Hogan	unincorporated area	Calaveras Co., California
Stockton East Water District	New Melones	unincorporated area	Stanislaus Co., California
Sweetwater Authority	Loveland	Alpine	San Diego Co., California
Sweetwater Authority	Sweetwater	Spring Valley	San Diego Co., California
Tahoe City Public Utility District	No Reservoir		
U.S. Bureau of Reclamation	Lewiston	Lewiston	Trinity, California
U.S. Bureau of Reclamation	Shasta Lake	Shasta Lake	Shasta Count, California
U.S. Bureau of Reclamation	Trinity		Trinity, California
U.S. Bureau of Reclamation	Whiskeytown Lake	Clear Creek, 7 mi. west of Redding	Shasta County, California

(continued)

Respondents to the MTBE and other fuel-related contaminants questionnaire (continued)

Utility	Reservoir Name	City	County
Vista Irrigation District	Lake Henshaw	Santa Ysabel	San Diego Co., California

APPENDIX D

MTBE AND OTHER FUEL-RELATED CONTAMINANTS QUESTIONNAIRE

2001

YOUR HELP IS NEEDED WITH THIS CRITICAL ISSUE!

The California Department of Health Services (DOHS) and the Awwa Research Foundation (AwwaRF) have awarded a research grant to the University of California at Davis, Contra Costa Water District (CCWD), and CH2M HILL. The purpose of the research is to model domestic water supply reservoirs containing untreated water to determine the impact on that water from MTBE and other fuel-related contaminants added by recreational and other watercraft. The research will culminate in a guidance manual that will educate utility managers, marina owners, boaters, public agencies, and others regarding the best ways to implement the project's research results. The manual and a user-friendly version of the model will be made available to you through AwwaRF upon completion of the project.

Instructions

Please forward this questionnaire to the appropriate person at your agency/utility.

The purpose of the questionnaire is to determine some basic facts about the reservoir(s) serving water to your utility. Please complete as much information as you can and return the form to:

CH2M HILL
Attn: Diana Sokolove
155 Grand Ave, Suite 1000
Oakland, CA 94612
dsokolov@ch2m.com

Contact Information

Please attach a business card for the person(s) that should be the primary contact for your utility or complete the information below.

Name & Title:____

Name of Utility:______________________________________

Address:____

City & Zip Code:______________________________________

Telephone:____

Fax: ____

E-mail: ____

Basic Reservoir Information

Please answer the following questions to provide us with basic information about your reservoir(s). If you have more than one reservoir, please complete a separate form for each waterbody.

Name of Reservoir:____________________________________

Location, City & County:____

Are motorized watercraft allowed? Yes_____No_____

If motorized watercraft are NOT allowed, approximately what year were they banned?

_________ Never allowed? Yes_____No_____

If motorized watercraft are NOT allowed, is this because of water quality concerns?

Yes_____No_____

If motorized watercraft are NOT allowed you do not need to complete the remainder of this questionnaire. Thank you for your assistance. If they are allowed, please continue the questionnaire on the next page.

Are two-stroke engines allowed? Yes_____No_____

Are personalized watercraft (e.g., jetskis) allowed? Yes_____No_____

Are the watercrafts licensed or permitted by your agency? Yes_____No_____

Does your agency or an authorized concessionaire rent motorized watercraft? Yes_____No_____

Are personal, versus utility owned, watercraft allowed? Yes_____No_____

Is there a patrol function by your agency? Yes_____No_____

Is there a limit on the time (day/hours) when they may be used? Yes_____No_____

How many days/week may they be used?______

Is the use allowed year-round? Yes_____No_____

Is there a limit on the number of watercraft allowed? Yes_____No_____

If YES, how many boats (approximately) are turned away?

______ daily / weekly (please circle one)

Are boating activities allowed throughout the waterbody (as opposed to being confined to a restricted area)? Yes_____No_____

Are there speed limits or other types of boating restrictions? Yes_____No_____

Details?____

____

Does the reservoir have a fueling station? Yes_____No_____

If a fueling station exists, does the fuel contain MTBE? Yes_____No_____

Do you, or any other group, keep records or statistics on watercraft use? Yes_____No_____

Can you provide approximate watercraft engine distribution information on the different types of watercrafts that use your waterbody?

______% two-stroke engines

______% four-stroke engines

______% non-powered

Can you provide information on the different types of activities at your waterbody? Please check all that apply.

______ fishing ______ jet skis/wave runners ______water skiing

______ other (please describe________________________________)

Can you provide any additional details about your past, present, and/or future control strategy(ies) for motorized watercraft on the reservoir? If you have written material that can be included with this questionnaire it will be appreciated.

____

____

____

____

____

____

____

Can you assess the level of public and/or governing board concern regarding MTBE and motorized watercraft associated with your waterbody?

None_____ Low_____ Medium_____ High ______

Physical Details

This information will be used to determine some basic facts about the hydrology of your reservoir.

Approximate surface area of the reservoir in acres?________

Approximate volume in acre-feet?________

Annual range?__________

Source of water for the reservoir? Please check all that apply.

______ local runoff ______ imported water ______upstream reservoir ______ other (please describe____________________________________)

Can you briefly describe the type of development immediately around the lake and in the area that drains directly to the lake (if applicable). We are interested in potential sources of urban pollution.

____

____

____

____

____

____

Are there private boat docks which provide access to the waterbody? Yes_____No_____

Maximum depth at normal operating range: _________________________

The user-friendly model will require certain inputs. Which of the following data do you collect? Please check all that apply.

______ wind data ______ water temperature ______relative humidity ______ air temperature

______ solar radiation

Water Quality Information

This information is meant to determine if water quality data exist for this reservoir.

Do you collect water quality data for MTBE for your reservoir? Yes_____No_____

Is there a regular monitoring program for MTBE? Yes_____No_____

If YES, please describe the program briefly (e.g., # samples/yr, # locations). If historic monitoring data are available and you are able to share it with the research team, please attach it to the questionnaire.

____

____

____

Are there reports or other information available regarding the limnology, water quality and/or operation of the waterbody? Yes_____No_____

The following two questions apply only to agencies/utilities that are also purveyors of finished drinking water:

Has MTBE been detected in the finished drinking water supply served by this reservoir?

Yes_____No_____

If you have historical finished water data on MTBE and are able to share it with the research team, please attach it to the questionnaire.

If MTBE in finished drinking water is of concern to you, have you considered changing or upgrading your water treatment process to deal with the MTBE problem? Yes_____No_____

Details?____

____

Additional Comments

____

____

____

____

____

____

____

____

____

____

____

____

____

Thank you for completing this questionnaire. If you have any questions, please contact John E. Reuter at (530) 304-1473, Fax (530) 753-8407, or jereuter@ucdavis.edu.

APPENDIX E

SUMMARY OF SURVEY FINDINGS

Table E.1

Level of concern over MTBE and reflection on control strategies

Reservoir Name	Utility and County	Comments
Level of Concern: High		
Calero	Santa Clara Water District, Santa Clara County	Two-stroke engines and personalized watercraft are allowed. Majority of the engines are two-stroke. The reservoir does not have a fuelling station. There is a regular MTBE monitoring program and MTBE has not been detected in the finished drinking water supply. Control strategies: limit the speed and number of water-crafts based on water quality; do not allow fuelling on the water; boating is not allowed throughout the reservoir.
Lake Arrowhead	Arrowhead Lake Association, San Bernardino	Two-stroke engines are allowed but not personalized watercraft. Majority of the engines are four-stroke. The reservoir has a fuelling station and the fuel does not contain MTBE. There is a regular MTBE monitoring program and MTBE has been detected in the finished drinking water supply. This is a private lake and only property owners in the adjacent tracks have access. Control strategies: limit property owners to number of boats, and number of slips; in the summer of 2000, only MTBE-free gasoline was provided at the gas dock and boaters supported this action paying more per gallon of gas; boating is allowed throughout the reservoir.

(continued)

Table E.1 (continued)

Reservoir Name	Utility and County	Comments
Anderson	Santa Clara Water District, Santa Clara County	Two-stroke engines are allowed but not personalized watercraft. Majority of the engines are two-stroke. There is a regular MTBE monitoring program and MTBE has not been detected in the finished drinking water supply. The reservoir does not have a fuelling station. Control strategies: PWC are no longer allowed due to concern about MTBE; limits on number of boats based on water quality; no fuelling allowed on the water; boaters are required to show a receipt indicating use of MTBE-free gasoline; boating is not allowed throughout the reservoir.
Salinas	County of San Luis Obispo, San Luis Obispo County	Two-stroke engines and personalized watercraft are allowed year round. Majority of the engines are two-stroke. There is no regular monitoring program for MTBE but data on MTBE are collected occasionally. MTBE was detected in the marina waters but not in the finished drinking water supply. The reservoir has a fuelling station and the fuel contains MTBE. Control strategies: There is no limit on number of watercraft; limit exists on day/hours used and activities are not allowed throughout the body. No additional information was provided on control strategies.

(continued)

Table E.1 (continued)

Reservoir Name	Utility and County	Comments
Palmdale Lake	Palmdale Water District, Los Angeles County	Two-stroke engines are allowed but not PWCs. Activities are allowed throughout the reservoir. No information was available about the distribution of engines. There is a regular MTBE monitoring program and during the fishing season the frequency of sampling is increased from monthly to weekly. In addition, from May 30, 2000 to July 14, 2000 the treatment plant influent, and effluent and clearwell effluent was also sampled, thrice a week. MTBE has been detected in the finished drinking water supply. The reservoir does not have a fuelling station. Control strategies: Limits on number of days / hours used, types of boats and speed with a speed limit of 10 mph; thee is no plan to upgrade the water treatment process for MTBE removal.
Littlerock Dam	Palmdale Water District, Los Angeles County	Two-stroke engines are allowed but not PWCs. Activities are allowed throughout the reservoir. No information was available about the distribution of engines. There is a regular MTBE monitoring program where samples were collected once a year in 1999. In 2000, samples were taken whenever water draw from the dam was started. MTBE has not been detected in the finished drinking water supply. The reservoir does not have a fuelling station. The US Forest Service supervises the recreation facilities at the Dam. Control strategies: Limits on number of days / hours used, and speed with a limit of 5 mph.

(continued)

Table E.1 (continued)

Reservoir Name	Utility and County	Comments
Lake Almanor	LACC Mutual Water Co., Plumas County	Two-stroke engines and PWCs are allowed throughout the reservoir. There are no limits on days / hours of usage. No information was available about the distribution of engines. The utility does not collect water quality data for MTBE, but the Department of Water Resources has a regular MTBE monitoring program. MTBE has not been detected in the finished drinking water supply. The reservoir has a fuelling station and the fuel contains MTBE. Control strategies: Limits on speed 5 mph at night and near shore; source of domestic water supply is four wells located near the lake, no water is taken from the lake for domestic purposes.

(continued)

Table E.1 (continued)

Reservoir Name	Utility and County	Comments
Railroad Canyon Reservoir	Elsinore Valley Municipal Water District, Riverside County	Two-stroke engines are allowed throughout the reservoir but not PWCs. There are no limits on days / hours of usage or number of crafts. 95 percent of the engines are two-stroke. MTBE data are collected and there is a regular MTBE monitoring program where a minimum of six samples per month are collected. The reservoir has a fuelling station and the fuel contains MTBE. MTBE has not been detected in the finished drinking water supply. Control strategies: Limits on speed; water treatment system is proposed to be upgraded by installing an Advanced Oxidation Process at the treatment plant to remove MTBE from the water; proposed to switch to no MTBE fuel; provide encouragement for conversion to four-stroke engines; work with local POA to educate residents.

(continued)

Table E.1 (continued)

Reservoir Name	Utility and County	Comments
Sweetwater	City of San Diego Water Department, San Diego County	Watercraft are allowed but not PWCs. The only watercraft that are allowed are those belonging to the Sweetwater Authority and they are used for reservoir monitoring and operational maintenance. The utility collects MTBE data and has a regular MTBE monitoring program. Sampling is done once a year at one location. The utility is also involved in a USGS monitoring program under which samples are collected fm seven locations on a quarterly basis and analyzed for VOCs by the USGS National Water Quality Laboratory. Only four-stroke engine boats are used and launched only when needed. The reservoir does not have a fuelling station. MTBE has not been detected in the finished drinking water supply. Control strategies: No recreational crafts are allowed.
Loveland	Sweetwater Authority, San Diego County	Watercraft are allowed but not PWCs. The utility collects MTBE data and has a regular MTBE monitoring program. Sampling is done once a year at one location. The utility is also involved in a USGS monitoring program under which samples are collected from seven locations on a quarterly basis and analyzed for VOCs by the USGS National Water Quality Laboratory. Four-stroke engines predominate. The reservoir does not have a fuelling station. No information was provided about MTBE occurrence in the finished drinking water supply. Control strategies: No PWCs are allowed; the only activities are jet skiing and routine patrolling using four-stroke engines; shoreline trash is picked up using two-stroke motors.

(continued)

Table E.1 (continued)

Reservoir Name	Utility and County	Comments
Lake Tahoe	North Tahoe Public Utility District, Washoe, El Dorado, Placer County	Watercraft and PWCs are allowed. Two-stroke engines are not allowed throughout the reservoir as of June 1999. There are no limits on days / hours of usage of crafts. MTBE data are collected and there is a regular MTBE monitoring program where a minimum of six samples per month are collected. The reservoir has a fuelling station and the fuel does not contain MTBE. The utility collects MTBE data and there is a regular annual MTBE monitoring program. MTBE has not been detected in the finished drinking water supply. Control strategies: Limits on engine types. After October 1, 2001, any engine that does not meet the USEPA's 2006 or the CARB 2001 emission standard will not be allowed; Speed limits apply; the utility extended the intake to a location off shore since 1995; upgradation of the treatment plant is being considered.
Level of Concern: Medium		
	Santa Clara Water District, Santa Clara County	PWCs and two-stroke engines are allowed all year round. However, these activities are not allowed throughout the year. Two-stroke engines predominate. The reservoir does not have a fuelling station. MTBE data are collected through a regular monitoring program where samples are collected monthly. MTBE has not been detected in the finished drinking water supply. Control strategies: Restrictions on speed and number of crafts based on water quality; fueling is not allowed on the water; boaters are required to show a recent receipt from one of the four local gasoline stations selling gasoline produced without MTBE to get on the water.

(continued)

Table E.1 (continued)

Reservoir Name	Utility and County	Comments
Lake Nacimiento	Nacimiento Water Company, Monterey County	Watercraft, PWCs and two-stroke engines allowed. They are allowed throughout the lake/reservoir and throughout the year. The reservoir has a fuel station and the fuel contains MTBE. Distribution of engines is almost equal between two-stroke and four-stroke. MTBE data are collected and there is a regular monitoring program. Samples are collected annually from wells downstream of the lake influenced by the lake. Detection of MTBE will lead to quarterly sampling. MTBE has not been detected in the finished drinking water supply. Control: strategies: Limits on speed with a 5 mile zone towards the narrow end; banning of two-stroke engines is being considered; two more filtration stages and ozonization have been added to the treatment process.
Ruth Lake	Humboldt Bay Municipal Water District, Trinity County	Watercraft, PWCs and two-stroke engines are allowed throughout the year and across the reservoir. Equal numbers of two- and four-stroke engines operate. The reservoir has a fuelling station and the fuel has MTBE. MTBE data are not collected and the reservoir is not part of a regular MTBE monitoring program. The utility stated that this was because MTBE was not detected in any of the previous samples collected by the Regional Water Quality Control Board. Control strategies: Enforcement of State boating law and speed limits near swimming areas.

(continued)

Table E.1 (continued)

Reservoir Name	Utility and County	Comments
Lake Hemet	Lake Hemet Municipal Water, Riverside County	Watercraft with two-stroke engines are allowed all through the year across the reservoir. PWCs are not allowed. Two-stroke engines dominate followed by non-powered engines. The reservoir does not have a fuelling station. MTBE data are collected but there is no regular MTBE monitoring program. MTBE has not been detected in the finished drinking water supply. The reservoir does not have a fuelling station. Control strategies: Speed limits with a restriction of 15 mph.
Lopez Lake	Lopez Lake- San Luis Obispo Co., San Luis Obispo County,	Watercraft and PWCs are allowed the year around throughout the lake/reservoir. Two-stroke engines are allowed. Two-stroke engines predominate followed by four-stroke and non-powered engines. The reservoir has a fuelling station and the fuel contains MTBE. The utility collects MTBE data, but there is no regular MTBE monitoring program. MTBE has not been detected in the finished drinking water supply. Control strategies: Speed limits with designated "No Wake" zones throughout the lake/reservoir; boating activities are not allowed within a 25 to 50 foot restricted area around the intake structure; no future control strategies are anticipated.

(continued)

Table E.1 (continued)

Reservoir Name	Utility and County	Comments
Lake Mary	Mammoth Community Water District, Mono County	Watercraft and two-stroke engines are allowed the year around throughout the lake/reservoir. PWCs are not allowed. No information was available on the distribution of engines operating on the reservoir. The reservoir has a fuelling station and the fuel contains MTBE. The utility collects MTBE data, and there is a regular MTBE monitoring program. MTBE has not been detected in the finished drinking water supply. Control strategies: Speed limits of 5 mph; the utility may wish to limit use of motorized watercraft due to potential for contamination.
Clear Lake	City of Lakeport, Lake County	Watercraft and PWCs, are allowed the year around throughout the lake/reservoir. Two-stroke engines are allowed. No information was available on the distribution of engines operating on the reservoir. The reservoir has a fuelling station and the fuel contains MTBE. The utility collects MTBE data, and there is a regular MTBE monitoring program. One sample per year is taken and the data reported to the California DHS. MTBE has not been detected in the finished drinking water supply. Control strategies: Speed limits with restrictions in areas close to the shore; the utility does not have any future control strategies at this time; the utility states that existing surface water treatment facilities are capable of treating MTBE if and when it is present in detectable levels.

(continued)

Table E.1 (continued)

Reservoir Name	Utility and County	Comments
Combie	Nevada Irrigation District, Placer and Nevada County	Watercraft and PWCs, are allowed the year around but not throughout the lake/reservoir. Two-stroke engines are allowed. No information was available on the distribution of engines operating on the reservoir. The reservoir does not have a fuelling station. The utility collects MTBE data, but there is no regular MTBE monitoring program. MTBE has not been detected in the finished drinking water supply. Control strategies: There is minimal watercraft activity on the reservoir due to surrounding private property ownership above the high water level currently. This has resulted in watercraft access by invitation only.
Scotts Flat	Nevada Irrigation District, Nevada County	Watercraft are allowed the year around and throughout the lake/reservoir but not PWCs. Two-stroke engines are allowed. Equal number of two-stroke and four-stroke engines operate on the reservoir. The reservoir has a fuelling station and the fuel contains MTBE. The utility collects MTBE data, but there is no regular MTBE monitoring program. MTBE has not been detected in the finished drinking water supply. Control strategies: Speed limits of 5 mph in the three "no wake" zones; regulations of the California Department of Boating and Waterways and County of Nevada are adopted.

(continued)

Table E.1 (continued)

Reservoir Name	Utility and County	Comments
Rollins	Nevada Irrigation District, Placer and Nevada County	Watercraft and PWCs are allowed the year around and throughout the lake/reservoir. Two-stroke engines are allowed. Equal numbers of two-stroke and four-stroke engines operate on the reservoir. The reservoir has a fuelling station and the fuel contains MTBE. The utility collects MTBE data, but there is no regular MTBE monitoring program. MTBE has not been detected in the finished drinking water supply. Control strategies: Speed limits of 5 mph in the three "no wake" zones; regulations of the California Department of Boating and Waterways and Counties of Nevada and Placer are adopted.
Jackson Meadows	Nevada Irrigation District, Sierra and Nevada County	Watercraft and PWCs are allowed the year around and throughout the lake/reservoir. Two-stroke engines are allowed. Equal numbers of two-stroke and four-stroke engines operate on the reservoir. The reservoir does not have a fuelling station. The utility does not collect MTBE data, and it does not have a regular MTBE monitoring program. MTBE has not been detected in the finished drinking water supply. Control strategies: No specific control strategies were reported during the survey.

(continued)

Table E.1 (continued)

Reservoir Name	Utility and County	Comments
French	Nevada Irrigation District, Nevada County	Watercraft and PWCs are allowed the year around and throughout the lake/reservoir. Two-stroke engines are allowed. No information is available about the distribution of types of engines that operate on the reservoir. The reservoir does not have a fuelling station. The utility does not collect MTBE data, and it does not have a regular MTBE monitoring program. MTBE has not been detected in the finished drinking water supply. Control strategies: No specific control strategies were reported during the survey.
Jackson	Nevada Irrigation District, Nevada County	Watercraft and PWCs are allowed the year around and throughout the lake/reservoir. Two-stroke engines are allowed. No information is available about the distribution of types of engines that operate on the reservoir. The reservoir does not have a fuelling station. The utility does not collect MTBE data, and it does not have a regular MTBE monitoring program. MTBE has not been detected in the finished drinking water supply. Control strategies: No specific control strategies were reported during the survey.

(continued)

Table E.1 (continued)

Reservoir Name	Utility and County	Comments
Sawmill	Nevada Irrigation District, Nevada County	Watercraft and PWCs are allowed the year around and throughout the lake/reservoir. Two-stroke engines are allowed. No information is available about the distribution of types of engines that operate on the reservoir. The reservoir does not have a fuelling station. The utility does not collect MTBE data, and it does not have a regular MTBE monitoring program. MTBE has not been detected in the finished drinking water supply. Control strategies: No specific control strategies were reported during the survey.
Faucherie	Nevada Irrigation District, Nevada County	Watercraft and PWCs are allowed the year around and throughout the lake/reservoir. Two-stroke engines are allowed. No information is available about the distribution of types of engines that operate on the reservoir. The reservoir does not have a fuelling station. The utility does not collect MTBE data, and it does not have a regular MTBE monitoring program. MTBE has not been detected in the finished drinking water supply. Control strategies: No specific control strategies were reported during the survey.

(continued)

Table E.1 (continued)

Reservoir Name	Utility and County	Comments
Bowman	Nevada Irrigation District, Nevada County	Watercraft and PWCs are allowed the year around and throughout the lake/reservoir. Two-stroke engines are allowed. No information is available about the distribution of types of engines that operate on the reservoir. The reservoir does not have a fuelling station. The utility does not collect MTBE data, and it does not have a regular MTBE monitoring program. MTBE has not been detected in the finished drinking water supply. Control strategies: No specific control strategies were reported during the survey.
Lake Casitas	Casitas Municipal Water District	Watercraft and PWCs are allowed the year around and throughout the lake/reservoir. Two-stroke engines are allowed. Most of the engines that operate on the reservoir are two-stroke engines. The reservoir has a fuelling station and the fuel contains MTBE. The utility collects MTBE data, and it has a regular MTBE monitoring program. MTBE has been detected in the finished drinking water supply but below the reporting detection limit. Control strategies: Presently there are no strategies for controlling or reducing motorized watercraft. The utility avoids using the surface water where the highest concentration of MTBE is found by selecting the appropriate intake level. An aeration system near the intake operates from April through November. During fall and winter when the lake is mixing, the intake structure becomes ineffective since the MTBE sinks to the lower depths.

(continued)

Table E.1 (continued)

Reservoir Name	Utility and County	Comments
Level of Concern: Low		
Shasta Lake	U.S. Bureau of Reclamation, Shasta County	Watercraft and PWCs are allowed the year around and throughout the lake/reservoir. Two-stroke engines are allowed. There are equal numbers of two-stroke and four-stroke engines that operate on the reservoir. The reservoir has a fuelling station and the fuel contains MTBE. The utility collects MTBE data, and it has a regular MTBE monitoring program. As part of the monitoring program, the lake is monitored at two locations before and after the Memorial day, July 4th and Labor day. Control strategies: No specific control strategies were reported during the survey.
Trinity	U.S. Bureau of Reclamation, Trinity County	Watercraft and PWCs are allowed the year around and throughout the lake/reservoir. Two-stroke engines are allowed. There are equal numbers of two-stroke and four-stroke engines that operate on the reservoir. The reservoir has a fuelling station and the fuel contains MTBE. The utility collects MTBE data, and it has a regular MTBE monitoring program. As part of the monitoring program, the lake is monitored at two locations before and after the Memorial day, July 4th and Labor day. Control strategies: Speed limits are imposed at night and in the marina. No other specific control strategies were reported during the survey.

(continued)

Table E.1 (continued)

Reservoir Name	Utility and County	Comments
Lake Henshaw	Vista Irrigation District, San Diego County	Watercraft and two-stroke engines are allowed the year around and throughout the lake/reservoir. PWCs are not allowed. All the engines that operate on the reservoir are four-stroke engines. The reservoir does not have a fuelling station. The utility collects MTBE data, and it has a regular MTBE monitoring program. As part of the monitoring program, two samples per year are collected from "mid-lake." These samples collected for three years (since 1997) have shown non-detectable levels of MTBE till the date of the survey. Control strategies: Since there are no PWCs and two-stroke fuelling facilities, MTBE has not been a concern. The lake is used for fishing crafts only. In addition, there are no gas stations within 15-20 miles of the lake. The utility reports that there is very good water quality control for VOCs.
Long Valley Reservoir (Crowley Lake)	Los Angeles Department of Water and Power, Mono County	Watercraft and PWCs are allowed the year around and throughout the lake/reservoir. Two-stroke engines are allowed. Two-stroke engines predominate on the reservoir. The reservoir has a fuelling station and the fuel contains MTBE. The utility collects MTBE data, and it has a regular MTBE monitoring program. A special program for Metropolitan Water District is being conducted for one year. One sample is collected every quarter when the water is not frozen at one location. MTBE has not been detected in the finished drinking water supply. Control strategies: The utility reported that there was no strategy at the time of the survey.

(continued)

Table E.1 (continued)

Reservoir Name	Utility and County	Comments
El Capitan	City of San Diego Water Department, San Diego County	Watercraft and PWCs are allowed the year around and throughout the lake/reservoir. Two-stroke engines are allowed. No information is available regarding the distribution of engines operating on the reservoir. The reservoir does not have a fuelling station. The utility collects MTBE data, and it has a regular MTBE monitoring program. As part of the monitoring program, quarterly sampling is carried out at surface and outlet depths. MTBE has been detected in the finished drinking water supply. Control strategies: No specific control strategies were reported during the survey.
Morena	City of San Diego Water Department, San Diego County	Watercraft and two-stroke engines are allowed the year around and throughout the lake/reservoir. PWCs are not allowed. About fifty percent of the engines operating on the reservoir are four-stroke. The reservoir does not have a fuelling station. The utility collects MTBE data, and it has a regular MTBE monitoring program. As part of the monitoring program, quarterly sampling is carried out at one station. MTBE has not been detected in the finished drinking water supply. Control strategies: No specific control strategies were reported during the survey.

(continued)

Table E.1 (continued)

Reservoir Name	Utility and County	Comments
Miramar	City of San Diego Water Department, San Diego County	Watercraft and two-stroke engines are allowed throughout the lake/reservoir. PWCs are not allowed. Two-stroke outboard engines predominate on the reservoir. The reservoir does not have a fuelling station. The utility collects MTBE data, and it has a regular MTBE monitoring program. As part of the monitoring program, quarterly sampling is carried out from the surface and outlet depths. MTBE has been detected in the finished drinking water supply. Control strategies: Boats are allowed only for four days in a week. No other specific control strategies were reported during the survey.
Lower Otay	City of San Diego Water Department, San Diego County	Watercraft and two-stroke engines are allowed the year around and throughout the lake/reservoir. PWCs are not allowed. Two-stroke outboard engines predominate on the reservoir. The reservoir does not have a fuelling station. The utility collects MTBE data, and it has a regular MTBE monitoring program. As part of the monitoring program, quarterly sampling is carried out from the surface and available outlet depths. MTBE has been detected in the finished drinking water supply. Control strategies: Boats are allowed only 3 days a week. No other specific control strategies were reported during the survey.

(continued)

Table E.1 (continued)

Reservoir Name	Utility and County	Comments
San Vincente	City of San Diego Water Department, San Diego County	Watercraft and two-stroke engines are allowed the year around and throughout the lake/reservoir. PWCs are not allowed. four-stroke outboard engines predominate on the reservoir. The reservoir does not have a fuelling station. The utility collects MTBE data, and it has a regular MTBE monitoring program. As part of the monitoring program, quarterly sampling is carried out from the surface and available outlet depths. MTBE has been detected in the finished drinking water supply. Control strategies: Boats are allowed only 4 days a week. No other specific control strategies were reported during the survey.
Sutherland	City of San Diego Water Department, San Diego County	Watercraft and two-stroke engines are allowed the year around and throughout the lake/reservoir. PWCs are not allowed. Two-stroke outboard engines predominate on the reservoir. The reservoir does not have a fuelling station. The utility collects MTBE data, and it has a regular MTBE monitoring program. As part of the monitoring program, quarterly sampling is carried out at one surface location. MTBE has been detected in the finished drinking water supply. Control strategies: Boats are allowed only 3 days a week. No other specific control strategies were reported during the survey.

(continued)

Table E.1 (continued)

Reservoir Name	Utility and County	Comments
Murray	City of San Diego Water Department, San Diego County	Watercraft and two-stroke engines are allowed the year around and throughout the lake/reservoir. PWCs are not allowed. Two-stroke outboard engines predominate on the reservoir. The reservoir does not have a fuelling station. The utility collects MTBE data, and it has a regular MTBE monitoring program. As part of the monitoring program, quarterly sampling is carried out from the surface and available outlet depths. MTBE has been detected in the finished drinking water supply. Control strategies: Boats are allowed only 3 days a week. No other specific control strategies were reported during the survey.
Big Bear Lake	Big Bear Municipal Water District, San Bernardino	Watercraft, PWCs and two-stroke engines are allowed the year around and throughout the lake/reservoir. Two-stroke and four-stroke outboard engines have equal distribution. The reservoir has a fuelling station and the fuel contains MTBE. The utility does not collect MTBE data, and it does not have a regular MTBE monitoring program. No information was provided on MTBE presence in the finished drinking water supply. Control strategies: No specific control strategies were reported during the survey.

APPENDIX F
MARINE ENGINE TECHNOLOGIES

PWCs

Kawasaki is the manufacturer of the original stand-up PWC (Jet Ski) They have been around for 30 years. In 2003 they will be selling nine three-seater models, twelve two-seater models, and eight single-rider stand-up models. One of the three-passenger models (1100 STX-DI) and one of the two-passenger models (Ultra 130 DI) will be equipped with Ficht injection technology. Additionally they have introduced a four-stroke engine that is available in a three-seater model (STX 12-F). These two craft will meet USEPA 2006 standards for emissions. The remaining fleet is standard two-stroke design.

Honda Power products entered the PWC business in 2001 with a four-stroke PWC. In 2003, Honda will have three four-stroke PWCs available. Two of these craft will be three-seater models (Aqua Trax F-12 and F-12X), and one will carry two passengers (Aqua Trax R-12X). All of the Honda PWCs will meet USEPA 2006 emission standards.

Yamaha, the manufacturer of WaveRunner watercraft, has developed a four-stroke engine that will be included in its 2003 product line. This engine will power two of Yahama's three-seater models (FX140 and FX Cruiser). Yamaha will also continue to produce its electronic fuel injection (EFI) engine that will power one of the two-passenger craft. The EFI and four-stroke engines meet the USEPA 2006 emission standards. In addition to the clean engine technology, Yamaha will have six PWCs powered by conventional two-stroke engines. SeaDoo is selling 11 PWC models in 2003. There are two two-seater models, eight three-seater models, and one that seats four people. Four of the three-passenger models are equipped with Sea-Doo's new four-stroke engine (4 TEC), one has direct injection technology, and the remaining three use Sea-Doo's two-stroke Rotax engine. Both of the two-passenger craft as well as the four-passenger models are powered by direct injection technology. Both the four-stroke and Rotax Orbital direct injection engines meet USEPA 2006 emission standards. This provides clean engine options (either four-stroke or direct injection) for each size class of craft in the 2003 Sea-Doo fleet.

The clean technology craft, regardless of the number of passengers, cost, on average, approximately $1,000 more than comparably equipped boats with conventional two-stroke engines. This figure is based on manufacturers-suggested retail price. The future expansion of clean engine technology (direct injection two- and four-stroke) in the PWC market will likely depend on consumer demand. The major manufacturers have met their obligation to produce clean engine alternatives. Consumers will determine whether these engines become the standard for all models across the Hp range.

OUTBOARD ENGINES

Yamaha has four classes of engines that cover a large Hp range. They are selling standard two-stroke engine technology as well as OX66 and HPDI two-stroke technologies and four-stroke engines. The HPDI is a high pressure direct injection system and meets USEPA 2006 emissions requirement. There are three models of the HPDI engines available with Hp ranging from 150 to 250. The OX66 two-stroke engines cover an Hp range of 150 to 250. It is unclear whether these engines meet USEPA 2006 standards.

Yamaha is also selling conventional two-stroke engines (loop charged) ranging from 8 to 200 Hp and four-stroke engines ranging from 2.54 to 225 Hp. All of the four-stroke engines meet USEPA 2006 emission standards.

Mercury Marine will continue to sell conventional two-stroke, four-stroke, and two models of injected two-stroke engine technologies. The four-stroke engines are available in 4 to 225 Hp and meet USEPA 2006 emission standards. The injected two-stroke engines (the EFI and OPTIMAX) cover the higher Hp alternatives with a range of 135 to 225 and 150 to 250 for the OPTIMAX and EFI respectively. Mercury Marine has stated that they will discontinue the sale of conventional two-stroke engine technology in the U.S. and Canada by 2006. The Bluewater Network presented an award for "excellence in environmental engineering" to Mercury for their attention to environmental issues with respect to motorized boating.

Bombardier, the manufacturer of the Evinrude product line, is selling a series of outboard engines that utilize Ficht Ram Injection. This is direct injection technology that meets USEPA 2006 emission standards. The engines range from 75-250 Hp.

Honda continues to produce four-stroke outboard engines that meet the USEPA 2006 emissions standards. They have been selling four-stroke outboards since 1972. The company is now producing four-stroke outboards that range from 2 to 225 Hp.

APPENDIX G
BOAT RAMP DATABASE EXAMPLE

Ramp Database Example

DATE	Weekday 1 = Sun. 7 = Sat.	Weekend 1= wkend 0 = wkday	Time	Timeslot (9 = 9-11)	BOAT #	IN/OUT 1 = launch 2 = recover	PWC - carb. 1 = yes	PWC - fuel inject. 1 = yes	2-stroke carb. (HP)	2-stroke fuel inject. (HP)	4-stroke (HP)	INBOARD 1 = yes	ENGINE TYPE (1-8)	TROLLING (hours)	SKIING (hours)
06/17/00	7	1	10:04	9	CF0850EC	1	0	0	0	0	0	1	8	0	0
06/17/2000	7	1	9:36	9	CF3894NX	2	0	0	0	0	0	1	8	0	1.5
06/17/2000	7	1	10:48	9	CF6090HG	1	0	0	0	0	0	1	8	0	0
06/17/2000	7	1	9:21	9	CF6133NL	1	1	0	0	0	0	0	1	0	0
06/17/2000	7	1	10:33	9	CF8404HT	2	0	0	0	0	0	1	8	3	0
06/17/2000	7	1	9:21	9	CF8470PL	2	0	0	200	0	0	0	5	1.5	1.5
06/17/2000	7	1	10:04	9	NV4794KN	2	0	0	0	0	8	0	7	2	0
06/17/2000	7	1	9:07	9	NV9447KN	2	0	0	0	0	0	1	8	0	0
06/17/2000	7	1	11:45	11	CF0194KE	2	0	0	6	0	0	0	3	6	0
06/17/2000	7	1	11:16	11	CF3960NC	1	0	0	0	0	0	1	8	0	0
06/17/2000	7	1	11:02	11	CF5166PD	1	0	0	0	0	0	1	8	0	0
06/17/2000	7	1	11:31	11	CF8975HW	1	0	0	60	0	0	0	4	0	0
06/17/2000	7	1	12:00	11	CF9523GW	2	0	0	0	0	0	1	8	3	0
06/17/2000	7	1	11:16	11	NV7858KP	1	1	0	0	0	0	0	1	0	0
06/17/2000	7	1	14:09	13	CF3087PS	1	0	0	0	0	150	0	7	0	0
06/17/2000	7	1	13:12	13	CF7499KK	1	0	0	30	0	0	0	4	0	0
06/17/2000	7	1	15:07	15	CF7002HC	1	0	0	0	0	0	1	8	0	0
06/17/2000	7	1	15:07	15	CF8626JM	1	0	0	125	0	0	0	5	0	0
06/18/2000	1	1	8:24	7	CF6005JM	1	0	0	0	0	0	1	8	0	0
06/18/2000	1	1	9:21	9	CF2039FP	2	0	0	20	0	0	0	3	0	0
06/18/2000	1	1	11:45	11	CF1981PG	0	0	0	0	0	0	1	8	0	0
06/18/2000	1	1	11:45	11	CF4652CS	2	0	0	33	0	0	0	4	1	0
06/18/2000	1	1	11:31	11	CF6710HV	2	0	0	0	0	15	0	7	4	0
06/18/2000	1	1	11:31	11	CF7406GP	2	0	0	50	0	0	0	4	0.3	0
06/18/2000	1	1	11:45	11	CF9523GW	2	0	0	0	0	0	1	8	2	0
06/18/2000	1	1	12:14	11	NV9447KN	2	0	0	0	0	0	1	8	0	2
06/18/2000	1	1	12:14	11	OR9UZ	2	0	0	0	0	8	0	7	3	0
06/18/2000	1	1	13:12	13	CF0758FY	2	0	0	115	0	0	0	5	0	0

APPENDIX H
SAMPLING PROGRAM DESIGN/RESULTS

PURPOSE OF SAMPLING PROGRAM

The primary purpose of the monitoring program was to obtain field data to allow for verification of the DLM-MTBE/BTEX model during the 2000 boating season at Donner Lake. Prior to using the model in a predictive fashion for guidance purposes, it was necessary to determine if the calibration was accurate; i.e. the results mimicked actual ambient conditions. To do this, an extensive field monitoring program for MTBE and BTEX was conducted to provide data on a number of scales including seasonal, weekly, daily, synoptic and distribution with depth from surface to bottom. Figure 3.7 in Chapter 3 shows the sampling locations. Since the boating occurs principally between Memorial Day (5/29/00) and Labor Day (9/4/00), sampling was clustered within this time frame.

A further benefit from the monitoring program was that it provided a good comparison with 1997 when a similar comprehensive study of MTBE was done in Donner Lake (Reuter et al. 1998). This additional year of data allows conclusions to be made regarding impacts of management differences between these two years (use of MTBE-free fuel by boat concessionaires in 2000) as well as the influence possible difference between boating and weather conditions between these two years. As noted elsewhere in this report, the original DLM-MTBE model was calibrated and applied to Donner Lake using the 1997 data.

SAMPLING DESIGN

Site Selection: Site selection was facilitated by results from the 1997 Donner Lake monitoring. The sampling team's extensive experience at Donner Lake allowed them to chose representative stations with a very high degree of confidence. The 1997 results showed that MTBE was distributed evenly throughout much of the surface area of Donner Lake. On those dates when MTBE exceeded a concentration of 1 μg/L, the coefficient of variation (standard deviation ÷ mean) for measurements taken at three sampling locations from west to east was low and typically <ten percent. The lake is characterized by an oval-shaped shoreline with no

sheltered embayments or enclosed marinas. The long, east-west axis also aligns with the prevailing winds, contributing to the spatially well-mixed conditions.

Based on these findings, two primary sites were chosen for sampling that were also sampled in 1997. Station 4 was located at mid-lake over the deepest portion of the basin with a maximum depth of 66 m. Since this was considered a representative location, sampling for seasonal, weekly and daily changes in MTBE/BTEX was done at Station 4. Station 5, at a maximum depth of 33 m, was positioned 1.4 km from the eastern outlet, but still within the main body of Donner Lake. Station 5 was intended as a supplement to Station 4 to confirm previous findings of similar MTBE levels. It was visited monthly between May and October with two extra samplings in July and one extra sampling in September in association with the Independence Day and Labor Day holidays.

Eight sites were visited during the mid-August synoptic sampling. These sites were positioned around the perimeter of the lake, and while not along the shoreline proper, each was located in the nearshore region relative to the open-water Stations 4 and 5.

Sampling Depths: In 1997, sampling at Station 4 was done at nine individual depths from surface to bottom (0, 3, 6, 9, 12, 22, 32.5, 45 and 57.5 m). The results conclusively showed little migration of MTBE below 8 to 12 m, the depth of the seasonal thermocline. Since MTBE was found almost exclusively above the thermocline and in the epilimnion, our 2000 sampling design focused on this layer. Four depths were considered adequate to characterize the vertical distribution of MTBE/BTEX in Donner Lake:

- 2 meters below the surface
- 1 meter below the mid-epilimnion
- 2 meters below the thermocline
- mid-hypolimnion

In mid-summer, these corresponded to depth of approximately 2 m, 4 to 6 m, 12 to 14 m and 35 m below the surface at both Stations 4 and 5.

Schedule: The Donner Lake 2000 boating season sampling schedule was designed to capture the three primary temporal scales which the model was targeted to predict. These include (1) seasonal time scale, i.e. looking at changes in MTBE/BTEX as boating season progressed during its typical Memorial Day to Labor Day period, (2) weekly time scale, i.e. what happens to MTBE and BTEX during weekend boating activity and the intervening weekday period between, and (3) daily time scale, i.e. did concentrations vary with time of boating activity on a given day.

While the 1998 version of the DLM-MTBE Model was able to predict on each of these time scales (McCord and Schladow 1998), this "first-generation" model could only be validated for the annual-seasonal scale, because data for the other times scales were unavailable. Below, is presented a summary of the sampling schedule used in 2000. In total, 156 samples were taken and analyzed for MTBE and the BTEX compounds (benzene, toluene, ethylbenzene and xylene). Note that the sum of the data points available for analysis for the four categories presented below is 169. The reason for the discrepancy between this value and the actual number of samples collected lies in the fact that some of the collected samples were able to be analyzed as part of more than one of the categories.

Seasonal: Data to define seasonal changes in MTBE/BTEX during the boating period were done on 14 dates in 2000 (May 22, June 12, June 26, July 5, July 10, July 17, July 24, August 7, August 21, August 28, September 5, September 11, September 25 and October 23). These were scheduled to provide maximum coverage during the July-early September when boating is typically greatest.

Sampling Dates	Station #	Depths	Total
14	4	4	56
9	5	4	45

Sampling was designed to be done on the same day of the week as to avoid inter-week differences. Monday was chosen as the sampling day since it immediately proceeded the weekend boating activities. The only exceptions to this were when the Independence Day and Labor Day holidays fell on a Monday in which case sampling was done on Tuesday. The time of sampling was also consistent between approximately 10:00 and 13:00. A total of 101 data points was available for this analysis.

Weekly: This scale of sampling was done during 2 weeks, one high and one lower boating activity week; June 8-14 and August 10-18, respectively. Collections were made at Station 4 on Thursday, Saturday, Sunday Monday and Wednesday. Samples were taken at the same four depths as presented above. The weekly collection on June 12 was designed to also serve as the seasonal sample at Station 4. Sample typically occurred around mid-day. A total of 40 data points were available for this analysis.

Days	Station #	Depths	Total
5 (June)	4	4	20
5 (August)	4	4	20

Daily: To see if detectable changes could be measured over the course of a daily period, a sampling was done in mid-summer (August 12). This was a Saturday when heavier boating was expected. Four time periods were included to cover daily differences in type and magnitude of boating activity. These times included 07:00, 11:45, 14:50 and 18:55 and were based on visual observations made in previous years.

The early morning and early evening sampling consisted of 5 epilimnion samples (2, 4, 6, 8 and 10 m), a single site just below the thermocline (14 m) and the mid-hypolimnetic (35 m) depth. The two mid-day sampling were done 2 m and 6, i.e. only epilimnion depths are being sampled for this component. A total of 18 data points was available for this analysis.

Hour	Station #	Depths	Total
07:00	4	7	7
11:45	4	2	2
14:50	4	2	2
18:55	4	7	7

Synoptic: To test for around-the-lake differences, a synoptic sampling was done once on Sunday, August 13. As discussed above, previous Donner Lake sampling showed that differences were minimal at three mid-lake stations sampled in 1997; however, this sampling was designed to test this finding relative to more nearshore stations. Eight nearshore stations plus Stations 4 and 5 were selected and sampled at a depth of 2 m. This depth was chosen as representative of the higher epilimnetic concentrations for MTBE. A total of 10 data points was available for this analysis.

Dates	Sites	Depths	Total
August 13, 2000	10	1	10

METHODOLOGY

Field Collection and Sample Preparation: Sampling for MTBE in Donner Lake was conducted from either a non-motorized kayak or a 14-foot aluminum fishing boat powered by a 15 horsepower, 4 stroke outboard motor. To ensure no contamination from the motorized vessel, all sites were approached from down wind, and the boat was anchored on location.

Water samples were collected from discrete depths using a Kemmerer well sampler. The two inch diameter stainless tube with polyurethane end seals was lowered to depth in the open position, and tripped closed with a messenger. The 0.6 L volume was hauled to the surface and used to completely fill pre-acidified 40 mL amber glass vials with septa sealed tops. Lake water was poured directly into these vials using the spring loaded purge valve at the bottom of the Kemmerer sampler. They were completely filled to ensure that no head space existed. Collected samples were kept in a cooler on ice and directly transported via overnight courier to the CH2M HILL Applied Sciences Group laboratory (Corvallis, Oregon) for analysis.

In situ temperature profiles were obtained at each station using a RBR temperature and depth data logger. Surface reference temperatures were taken with a hand held thermometer at each sampling location.

Laboratory Analysis: This section provides specific operating procedures for analyzing volatile organic analyses at CH2M HILL's Corvallis Applied Sciences Laboratory (CVO ASL) and describes the analysis of VOCs from environmental water, soil, and sediment samples using capillary-column gas chromatography with purge-and-trap concentration and mass spectrometry detection (GC-MSD). Gas chromatography and quality control procedures are based on USEPA Method 8000B, revision 2.0. Purge-and-trap concentration procedures are based on USEPA Method 5030B, revision 2.0. Analysis procedures are based on USEPA Method 8260B, revision 2.0. All of these methods are published in USEPA SW846, Test Methods for Evaluation of Solid Waste, May, 1997.

Upon receipt at the laboratory, sample containers are inspected for their condition. Information is verified by sample custody personnel before samples are processed into the laboratory information management system (LIMS). Each sample is assigned an unique laboratory identification used for tracking the sample within the laboratory. Sample temperature and holding time allowances were met for this project each time coolers where shipped.

During the analysis process, the sample is spiked with the appropriate volumes of internal standard and surrogate solution, a 5-mL aliquot of sample is removed from the sample container by the autosampler, and transferred to the purge-and-trap concentrator. Here the volatile components are transferred from the aqueous phase to the vapor phase, and swept onto a sorbent trap in the purge-and-trap concentrator. The volatile components are then thermally desorbed from the trap and injected onto a gas chromatographic column. Once on the column, the gaseous components are separated and detected by the mass spectrometer detector (MSD). All chromatographic data are acquired using computerized operating software which is convenient for post-run processing, review, reporting, and data archival.

The instrument is calibrated regularly, and blanks, calibration verification standards, laboratory control samples, internal standards, surrogates, and other quality control measures are conducted routinely to ensure that the instrument is operating within proper parameters.

QA/QC

Field

Previous field QA/QC studies showed that field blanks (laboratory deionzed water passed through the Kemmerer sampler and other phases of the field collection process) were always below a 0.1 μg/L estimated limit of detection (LOD) and that duplicate samples were not significantly different (Reuter et al. 1998, Allen et al. 1998).

Laboratory

The CH2M Hill Applied Science Group laboratory has establish acceptance criteria for a number of QA/QC categories which must be met for the MTBE/BTEX analyses to be validly reported. These categories include calibration (<20 percent drift), blanks (less than the reporting level), duplicate samples (<20 relative percent difference), spike sample(s) (75 to 125 percent), lab control samples (75 to 125 percent) and surrogate recoveries. Specific chemical surrogates, along with their approximate control limits, included dibromofluoromethane (75 to 125 percent), 1,2-dichloroethane-d4 (75 to 115 percent), toluene-d8 (85 to 110 percent) and 4-bromofluorobenzene (85 to 115 percent). According to QA/QC reports which accompanied each set of results, acceptance criteria were met.

The reporting levels or analytical limits of detection were:

MTBE	0.2 µg/L
Benzene	0.2 µg/L
Toluene	0.2 µg/L
Ethylbenzene	0.2 µg/L
m,p – xylene	0.4 µg/L
o – xylene	0.2 µg/L

RESULTS

Frequency of Detection: Of the 156 total number of samples analyzed for MTBE and BTEX 89 (57 percent) contained MTBE at a level greater than or equal to the 0.2 µg/L detection limit. However, of the 88 samples taken in the epilimnion, 86 (98 percent), or nearly all of them contained MTBE at a level greater than or equal to the 0.2 µg/L detection limit. At no time did MTBE exceed the California secondary drinking water standard of 5 µg/L. The BTEX compounds were much less frequently found. Benzene rose above its 0.2 µg/L detection limit on only a single occasion when a concentration of 0.3 µg/L was measured a site (f) during the mid-August synoptic survey. This accompanied the highest single MTBE value found in this study of 3.6 µg/L. For comparison, the California primary drinking water standard for benzene is 1 µg/L.

Toluene was the most commonly measured BTEX compound found in 23 of the 88 eplimnetic samples (26 percent). Toluene typically ranged from 0.2 to 0.6 µg/L; however a single higher value of 1.5 µg/L was found, again at site (f) during the mid-August synoptic. Even the largest toluene concentration observed was much lower than the California primary drinking water standard of 150 µg/L and the USEPA taste and odor threshold of 42 µg/L. Similar to benzene, ethylbenzene was only found on a single occasion (again the synoptic survey, site f) but at the 0.2 μg/L detection limit. For comparison, the California primary drinking water standard is 700 μg/L while the USEPA taste and odor threshold is 29 μg/L.

Xylenes were found in only 3 of the 88 epilimnetic samples (3 to 4 percent). For two of these samples only o-xylene was seen (both at the 0.2 μg/L detection). The synoptic (site f) sample contained both m,p-xylene (0.9 μg/L) and o-xylene (0.4 μg/L). Again, these few observed concentrations were well below the California primary drinking water standard of 1,750 μg/L and the USEPA taste and odor threshold of 17 μg/L.

In their study of southern Californian drinking, surface lakes and reservoirs, Dale et al. (2000) found identical frequency of occurrence values to ours for MTBE, benzene, toluene and ethylbenzene. Their approximately 20 percent relative frequency for total xylenes was higher than our 3 to 4 percent value.

Distribution With Depth: The transport of MTBE to depth was significantly retarded by the thermocline. This was identical too the observation made in 1997 (Reuter et al. 1998). Between June 8 and October 23, this thermal barrier was strongly developed with the top of the thermocline, deepening as the summer progressed. In late May, the thermocline was beginning to develop at shallow depths. Since one of the sampling depths was consistent at 2 meters below the thermocline, the seasonal deepening did not affect the relative position of this site.

Thermocline Depth (m)

Date	May 22	Jun 26	Jul 17	Aug 26	Sep 11	Oct 23
Depth	2-3 m	8 m	9.25 m	11 m	14 m	18.75 m

The distribution of MTBE with depth was the same at Station 4 and Station 5. The confinement of MTBE to epilimnetic waters is obvious, with no MTBE found at a depth 2 meters below the thermocline. At both Stations 4 and 5, MTBE was found (above the 0.2 μg/L detection limit) on only two occasions. The first was on July 5th when MTBE at 12 m (3 m below the top of the thermocline) was 0.3 μg/L; just above the detection limit. The second was on October 23 when a concentration of 0.4 μg/L was measured at 35 m (mid-hypolimnion). The July instance was most likely do to a temporary entrainment while the October occurrence was most likely an anomalous point. Regardless, the data clearly show that the thermocline acted as an effective barrier to the downward movement of MTBE. None of the BTEX compounds were found below the thermocline.

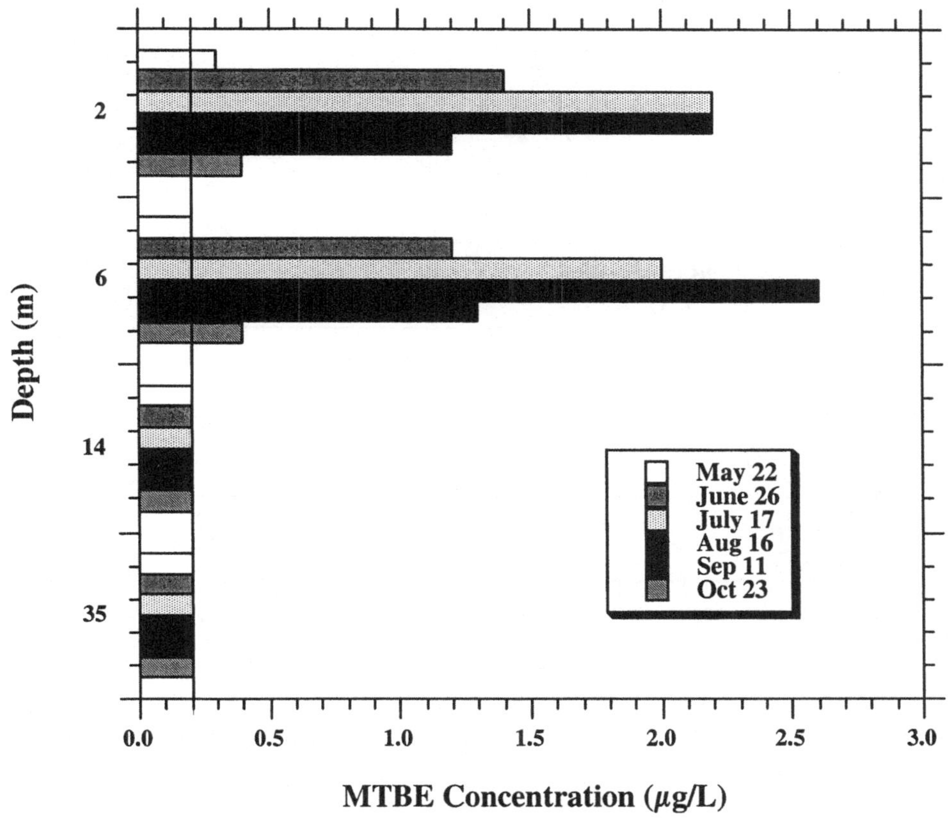

Figure H.1. Depth distribution of MTBE in Donner Lake. The 14 m depth represents a position in the water column 2 m below the thermocline. On September 11 the actual sampling depth was 16 m and on October 23 it was 21 m, reflecting the seasonal deepening of the thermocline.

Seasonal: A total of 23 individual sampling dates were available for construction of the seasonal changes in MTBE concentration (Figure H.2). These included the routine seasonal sampling plus the two intensive weekly sampling dates. Data are plotted as the mean epilimnetic level.

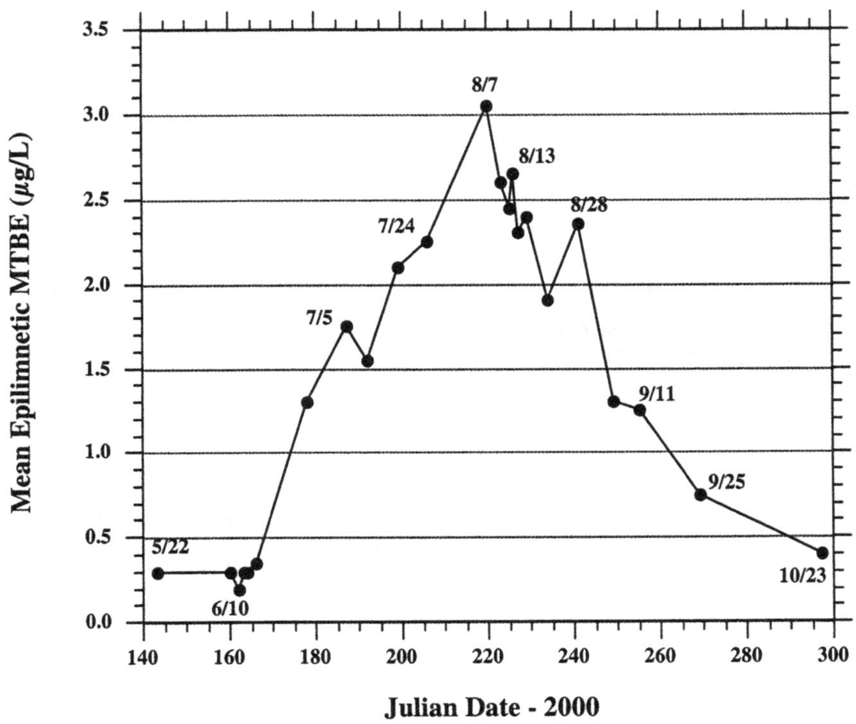

Figure H.2. Seasonal distribution of epilimnetic MTBE in Donner Lake, May 22 – October 23, 2000.

The distribution of MTBE in Donner Lake followed a clear cyclical pattern with a number of distinct periods. This pattern has been previously seen at a number of lakes and reservoirs including Lake Tahoe (Allen et al. 1998), Lake Perris in southern California (Dale et al. 2000) and the previous Donner Lake study (Reuter et al. 1998). During the 2000 Donner Lake study, MTBE was low, albeit above detection, from late-May to mid-June (0.2 to 0.4 μg/L). Beginning in late-June concentration began to steadily increase until August 7 when the maximum level of 3.1 μg/L was measured. This corresponded to the onset of boating. A significant difference between 1997 and 2000 was that in 1997 an 8 to 10 μg/L increase in MTBE was seen between July 1 to 7; associated with heavy boating over the Independence Day holiday. In 2000, no such spike in MTBE was seen and the absolute concentrations of MTBE were much lower (e.g. maximum MTBE in 1997 was 12.1 μg/L but only 3.1 μg/L in 2000.

During the month of August, 2000, MTBE fluctuated with a number on increase/decrease cycles; however, the period from August 7 through the end of sampling on October 23 was characterized by an overall decline in MTBE. It is noteworthy that in August, 1997 a general reduction was also observed amid the peaks; however, the drop was not as strong as in 2000.

No spike was observed during the Labor Day holiday and by mid-September MTBE declined to <1.5 μg/L. On September 25 concentration dipped below 1.0 μg/L and on the final sampling date (October 23) MTBE was at 0.4 μg/L and near the limit of detection. The 1997 study showed that by January at the latest, MTBE was <0.1 μg/L. The calculated half-life for MTBE based on the observed decline between August 28 and October 23 was 22 days. At this rate, MTBE would be expected to be <0.1 μg/L by mid-December.

Weekly: Sampling for weekly changes in MTBE was done on two occasions; mid-June when boating activity was low and mid-August when activity is seasonally high. The BTEX compounds were not above their detection limits for either of these periods; consequently, the presentation of data are limited to MTBE. Similarly, concentrations in mid-June were very low. Despite the fact that all the epilimnetic samples were above detection, the range and the absolute values was only 0.2 to 0.4 μg/L – too small to draw conclusions.

The mid-August data showed daily changes in MTBE concentration, albeit these were smaller than the more general seasonal changes. By adding a sampling profile done on the previous Monday to the Thursday-Wednesday sequence, differences are detected during the week (Figure H.3). On Monday August 7, the mean epilimnetic MTBE concentration at Station 4 was 3.05 μg/L and reflective of the preceding weekend boating activity. MTBE declined approximately 15 percent, to 2.60 μg/L, by Thursday (August 10). By Saturday morning at 07:00 MTBE had further declined to 2.45 μg/L or a 20 percent decrease relative to the previous Monday's high concentration. The observed drop in MTBE was most likely the result of minimum boating activity (new inputs) combined with loss due to volatility. The early Saturday sampling occurred before the start of the weekend boating. By Sunday, August 13, MTBE rose by just under 10 percent from 2.45 to 2.65 μg/L. Following the weekend, concentrations dipped to 2.30 μg/L on Monday and remained relatively stable on Wednesday (August 16). The 13 percent descent between Sunday and Monday was similar to the 15 percent drop between the previous Monday and Thursday.

Although the day-to-day changes detected during the week of August 7 to 16 were small and certainly not as dramatic as those seen in the seasonal record, some differences were evident and qualitatively explainable by boating activity. The largest change in MTBE was the 0.75 μg/L or 25 percent decrease from Monday to Monday during this week. As shown by the data defining seasonal changes (Figure H.2), the variation in MTBE between over the period August 7 to 16 are discernible on this plot.

The 1997 Donner Lake study did not attempt to resolve changes in MTBE over such a short time scale (i.e. weekly). However, for two sampling dates (July 1 and July 7) designed to assess loading over the Independence Day holiday, a very significant increase was observed in concentration from 2 μg/L to approximately 11 μg/L. This indicates that largely weekly changes are possible.

In the only published study found which investigated MTBE variation within a week, Baehr and Reilly (2001) reported from a 1999 study in Lake Hopatcong, NJ. They sampled five times between Aug 26 to September 3 and 4 time between September 10 to 20. MTBE was typically higher on Sundays and Mondays. The authors presumed this was due to increase weekend boat traffic. Absolute concentrations varied from 5.7 to 18.4 μg/L (at 3 foot depth) with increases over the weekends ranging from 1.2-2.7 times the initial value. Declines from Monday to Wednesday were 17 percent and 54 percent.

Daily: Monitoring for daily changes in MTBE was done on a mid-August weekend when boating activity was expected to be seasonally high. As presented above, sampling was done at four times between 07:00 and 18:55 to capture the diel pattern of boat use in Donner Lake. The analytical results suggest minor differences in MTBE concentrations on this time scale (at least on the day of sampling). The mean ± SD were 07:00 - 2.45±0.06 μg/L; 11:45 – 2.60±0.28 μg/L; 14:50 – 2.65±0.21 μg/L; and 18:55 – 2.63±0/13 μg/L. Clearly, MTBE concentrations between 11:45 and 18:55 are nearly identical at 2.60-2.65 μg/L. The 0:700 time was 0.15-0.20 μg/L lower and may have been reflective of less MTBE in the water at the start of the boating weekend. However, there was no statistically significant difference between 07:00 and 18:55 ($p=0.05$). Statistical differences for the other time periods are less meaningful since each time had only two replicates.

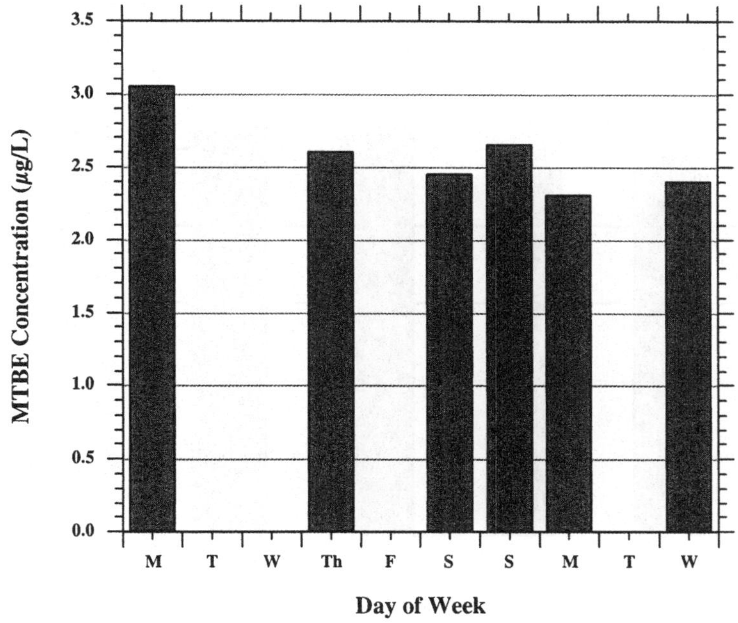

Figure H.3. Change in MTBE concentration at Station 4 over a 10-day period in mid-August when boating activity is seasonally high

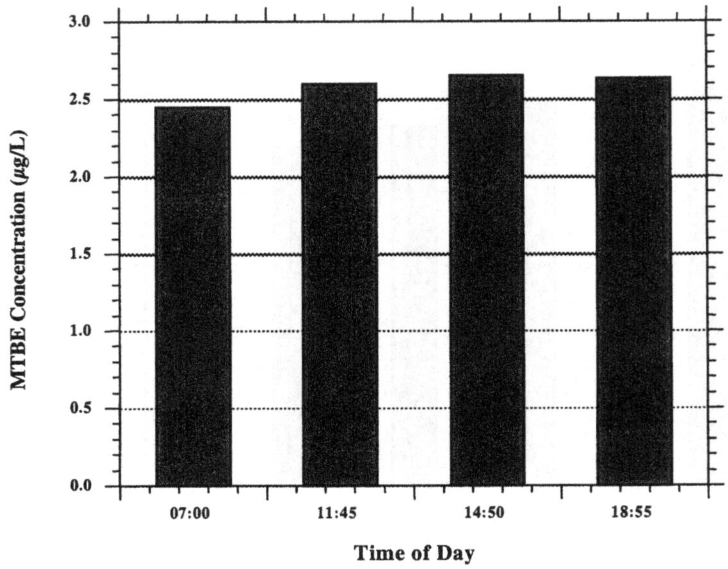

Figure H.4. Daily changes in MTBE in Donner Lake as measured on a mid-August weekend

Synoptic: As part of the effort to determine spatial variation between locations, Table H.1 presents a statistical comparison of the epilimnetic depths at Station 4 versus Station 5. There were 17 occasions when at least one of the MTBE concentrations was greater than the 0.2 μg/L limit of detection.

Table H.1

Comparison of epilimnetic MTBE concentrations (μg/L) at Stations 4 and 5.

Station #4	Station #5
0.3	0.4
1.4	1.4
1.2	1.1
1.4	1.6
2.1	1.6
1.7	1.9
1.4	1.9
2.2	1.9
2.0	1.6
2.9	3.5
3.2	2.8
1.2	1.2
1.4	1.2
0.6	0.9
0.9	>0.2
0.4	0.4
0.4	0.4
Mean ± Standard Deviation 1.5 ± 0.7	1.4 ± 0.8

On the basis of a two-tailed, paired t-test, there was no statistical difference between Stations 4 and 5 (p=0.63). Indeed, there was good agreement between the stations with a correlation coefficient (r) of 0.92. Higher concentrations of MTBE were found at Station 4 seven times while Station 5 was higher six times. The mean difference between these stations was 0.26 μg/L with a maximum difference of 0.5 to 0.7 μg/L. The greatest discrepancy was at a depth of 7 m on September 25 when a level of 0.9 μg/L was found at Station 4 with a below detection value (<2 μg/L) at Station 5.

The synoptic survey of Donner Lake was performed in mid-August, on a weekend when boating activity was expected to be elevated. The data presented below begin at site (a) near the Truckee Donner Recreation and Park District public boat ramp and circles the lake counter clockwise. The sampling began at 12:20 and ended approximately two hours later at 14:25. Both MTBE and toluene were above detection, thereby allowing for an analysis of each.

Table H.2

Synoptic sampling conducted on August 13, 2000

Location	MTBE	Toluene
a	2.8	0.4
b	2.8	0.3
c	2.8	0.2
d	2.6	0.4
e	3.0	0.6
f	3.6	1.5
g	2.4	0.3
h	2.5	0.2
#4	2.4	<0.2
#5	2.4	0.3
Mean ± SD	2.73 ±0.37	0.44 ± 0.39

Values expressed as μg/L

MTBE concentrations around the lake varied by a maximum of 1.2 μg/L during the synoptic survey. This was primarily the result of two sites (e) and (f) which were noticeably elevated relative to the remaining locations. These two sites had concentrations of 3.0 and 3.6 μg/L, respectively compared to the others which had a mean ± SD of 2.6 ± 0.2 μg/L. The coefficient of variation for these remaining locations is <10 percent indicting very little difference. Both sites (e) and (f) were positioned at the far west end of Donner Lake, a shallower area which boaters often use as a "destination" for cruising (picnicking on beaches) and the location of the PWC and boat concessionaire. Field notes on the day of sampling indicated high PWC use in this area (B. Allen unpub.).

The highest concentrations of toluene were found during the synoptic survey. Not unexpectedly, this also occurred at sites (e) and (f), 0.6 and 1.5 μg/L toluene, respectively. The remaining sites had a mean ± SD of 0.3 ± 0.1 μg/L; just above the 0.2 μg/L limit of detection. It is noteworthy that toluene was consistently found in the water during the synoptic survey even though MTBE was not higher. This result is due to the fact the synoptic sampling on a Sunday captured more recent or "fresher" marine engine emissions to the lake. Since the BTEX compounds are more volatile than MTBE, it is expected that the ratio of BTEX to MTBE would be maximum immediately after release and that this ratio would decline with time as the BTEX compounds are lost to the atmosphere. Site (f) was the only instance during the entire May-October monitoring program when MTBE and each of the four BTEX compounds were above detection at the same time.

In summary, both the synoptic sampling and the Station 4 versus Station 5 comparison indicate little spatial variation in MTBE and BTEX in Donner Lake, supporting previous conclusions (Reuter et al. 1998).

CONCLUSIONS

1) BTEX was found only infrequently in the Donner Lake, year 2000 monitoring. Values rarely exceeded 2-3 times their analytical limits of detection. At no time did BTEX approach a concern, vis-à-vis drinking water standards. Given the higher volatility of BTEX relative to MTBE, this was not surprising. The lack of BTEX in Donner Lake does not imply that it this is not an important fuel constituent. According to Fiore et al. (1998), toluene accounts for 8 percent of California Certified Fuel while total xylenes account for 11 percent (MTBE 14 percent). The toluene and total xylenes exhausted into the water from a 80 hp two-stroke carbureted PWC in Lake Tahoe and measured after 30 minutes of operation were 17 μg/L and 23 μg/L. Not only are these concentrations relatively high, their ratios to MTBE was similar to that in the raw fuel. These same authors reported toluene and xylene levels in an area of high PWC use in the range of 5-10 μg/L. Similarly Dale et al. (2000) found maximum toluene levels of 6 to 7 μg/L in Lake Havasu with xylene at about 3 μg/L. Because BTEX can be volatile, the elapsed time since discharge/loading will affect its concentration of surface lakes and reservoirs. Since sampling can not always be that responsive to boating activity or other sources, the use of a model may be the best way to predict expected BTEX levels. Even the frequent below detection date found at Donner Lake also for verification of the model.
2) In Donner Lake, the spatial distribution of MTBE was relatively uniform. This was especially true in the open waters. Along the shoreline, concentrations were slightly

higher in an area of extensive PWC use, but not more than 1 μg/L above the other nearshore sites. This substantiates our use of the 1-dimensional DLM-MTBE model. Lakes and reservoirs which irregular shorelines (e.g. bays, coves, arms) might be expected to demonstrate more spatial variation depending on size, volume, mixing patterns and currents.

3) As shown for a number of lakes and reservoirs, the summer thermocline in Donner Lake acted as an effective barrier to the downward transport of MTBE and BTEX to deeper depths. MTBE was rarely found below the thermocline, suggesting that modifying the depth of withdrawal for drinking water could be an effective management option, provided other conditions for water quality and treatment were met (e.g. low suspended solids, reduced taste and odor compounds, etc.). When the lake mixes in the fall/winter, MTBE will temporarily be transported to deeper depths.
4) The seasonal distribution of MTBE in Donner Lake followed a distinct cyclical pattern with a steady increase in mean epilimnetic concentration between the end of June and early August. Peak MTBE was on August 8 at a concentration of 3.1 μg/L. Elevated levels were not associated with either the Independence Day nor Labor Day holidays. Cold weather conditions at both these times resulted in reduced boating. The calculated half-life for MTBE at the end of the effective boating season was 22 days.
5) Differences in MTBE concentrations were detected over the course of a ten-day, mid-summer period. Levels during or immediately following the weekend were higher with 10 to 20 percent changes found over this period. Changes measured over this time scale were evident on the plot of seasonal patterns.

Daily changes in MTBE were not pronounced. The largest difference was found when MTBE rose from 2.45 μg/L at 07:00 to 2.65 μg/L at 14:50.

APPENDIX I
USER GUIDE TO SPREADSHEET MODEL

INTRODUCTION

This spreadsheet-based software estimates the maximum level of a contaminant in a reservoir during the boating season. It is based on a series of computer model runs using a reservoir model (DLM-MTBE) developed at the University of California, Davis. An intensive boating use survey was conducted at Donner Lake, California in 2000. The survey monitored the number of boats and their engine types, the activities in which they were engaged, length of time on the lake, and fuel use. As a result of the survey, boating use patterns could be related to contaminant levels in the lake. DLM-MTBE was developed and verified using the Donner survey data. DLM-MTBE operates on a daily cycle. It uses daily average weather and hydrologic data as input and computes vertical profiles of VOCs, temperature, and density for each day modeled.

This estimator is based on model runs that were made by keeping weather and hydrologic data constant for each day modeled during a particular run. Then, a set of runs was made for each environmental variable, varying it over the expected range of values. From these runs, regression curves were developed which related changes in the input variable to changes in VOC concentration in the water.

WEATHER DATA SOURCES

The estimator operates on average daily values for meteorological parameters. Ideally, weather data are measured at a point on or near the reservoir, but if such is not available, weather data may be obtained from other sources:

For the western U.S., contact the Western Region Climate Center:

http://www.wrcc.dri.edu/

(775) 674-7010

For all areas, contact the National Climatic Data Center:

http://lwf.ncdc.noaa.gov/oa/ncdc.html

http://lwf.ncdc.noaa.gov/oa/climate/climatedata.html

DATA ENTRY PRECISION

In general, the estimator will accept data more precise (more digits after the decimal point) than what is displayed on the entry sheet. The precision shown, however, is sufficient for the estimator's level of calculation, and additional precision will not change the estimate.

INPUT DATA: METEOROLOGICAL DATA

Average Daily High and Low temperatures

Enter the daily high and low temperatures in degrees Fahrenheit for your location. These two values are combined to estimate a daily average temperature. If the daily average is available, enter it in both boxes.

Relative Humidity

Enter the average daily relative humidity as a percentage between 0 and 100.

Cloud Cover

Check the appropriate button for average cloud cover during daylight hours. This information is used in calculating the amount of solar radiation reaching the reservoir surface.

Wind

Wind speed is the most important environmental variable affecting VOC levels in the reservoir. There are two ways to enter it. If a 24-hour average wind speed is available, enter it (in miles per hour) in the upper box and check the button to the left. When a 24-hour average is not available, check the button for average wind speed during the day and enter the estimated value in the first of the two boxes. In the second box, enter the number of hours during the day that the wind blows.

INPUT DATA: RESERVOIR PARAMETERS

Reservoir Area

Choose a reservoir height for the estimation, and enter the corresponding surface area in acres.

Reservoir Volume

Enter the volume in acre-feet for the same reservoir height as above.

INPUT DATA: BOATING USE INFORMATION

The number of boats and the types of engines they have determine the amount of fuel contaminants entering the reservoir.

Number of Boats

In the upper box, enter the total number of motorized boats (including PWC) using the reservoir each day. This will probably be a number between the average weekday use and average weekend use. If you have weekend and weekday numbers, you can multiply the weekday number by 4/7ths and the weekend number by 3/7ths and add the two results to obtain a daily average. The assumption behind this calculation is that there are 4 weekdays (Monday – Thursday) and three weekend days (Friday – Sunday).

Two-Stroke Engine Ban

Checking this box will remove all carbureted two-stroke engines from the calculation, and leave only fuel-injected two-strokes and four-strokes. It is a quick way to see the impact of conventional-technology two-stroke engines.

Internally, the estimator breaks down the number of boats into eight types of engines and three activities in order to calculate fuel consumption and emissions. It does this through information obtained in an extensive survey of boating activity at Donner Lake in 2000. You control the first part of the process by choosing to enter either percentages of engine types in the three overall categories, or by entering percentages of boats engaged in the three activities (fishing, skiing, and cruising). For example, if you say that 40 percent of your boating activity consists of PWCs, the estimator will split those PWCs into two types (carbureted and fuel injected) according to the numbers of each type seen at Donner in 2000.

Percent of Boats that are PWCs and Outboards

If you choose to break down boating activity according to the percentages of each major engine category, check the button for option one, and then enter the percent of your boat mix that are PWCs and outboards. The program will then calculate the percentage of inboards (including inboard/outboards) such that the total is 100 percent.

Percent of Boats Engaged in Fishing and Skiing

If you wish to break down your boating by the type of activity, check the button for option 2, and enter the percent of boats engaged in fishing and skiing. The program will calculate the percentage of cruising boats.

ASSUMPTIONS

A number of assumptions are built into the estimator. These include:

- All fuel used contains MTBE
- In places where you do not enter boating information, the patterns encountered at Donner Lake in 2000 are used. When you elect to enter percentages of engine types (boating option 1), for example, the estimator uses those percentages. But to split up PWCs into carbureted and injected, it uses the Donner percentages of each. Also, when you choose boating option 1, you do not enter boating activity percentages, and the Donner values are used.
- Internal tables containing average length of time spent on the lake, and average fuel consumption per hour for each activity and engine type, are constructed with values from the Donner survey.

APPENDIX J
EVALUATION OF THE ECONOMIC IMPACTS OF RECREATIONAL BOATING

INTRODUCTION

To fully understand the economic impact (on any particular lake or reservoir and region within which the lake or reservoir occurs) of the various alternatives discussed in this manual there is a need to quantify the current contribution of boating to the region. One possible way to quantify the economic contribution of boating is to undertake a regional economic impact analysis.

Regional economic impacts represent the changes in the economic activity within a region that result from an increase in the final demand for goods and services produced in the region. Any new spending in the region by persons and businesses located outside the region would be considered "exogenous" or as income coming from outside the region. Total economic impacts are composed of direct, indirect, and induced effects of the exogenous change in final demand. In the case of recreation, direct effects are represented by first-round purchases made by recreation visitors. Indirect effects occur as a result of purchases made by businesses to meet the increased demand for their products by recreation visitors. Direct and indirect effects lead to an overall production increase that can lead to more local or regional employment and income as other industries respond to meet the new demands of the directly affected industries. Induced impacts result from the increased spending by resident households whose incomes have increased as a result of the direct and indirect effects.

Regional economic impacts of recreation are typically assessed on the basis of visitor trip expenditures[1]. The money spent by visitors on food, lodging, and transportation is the input into the local economy. Management alternatives that impact/affect the amount or type of money spent will affect the local economy.

Typically, only the trip expenditures of nonresident visitors are considered when assessing the impact of recreation on a local economy. The primary reason for excluding local

[1] Expenditures on capital goods are not included because (1) these goods are mostly likely purchased in the visitors' home county /state (in which case none of that money finds its way into the local economy of the recreational area) and (2) there is no easy way of splitting the cost among the various recreation trip destinations.

residents trip expenditures is that these expenditures do not represent infusions of new dollars into the local economy.

ASSUMPTIONS

Input-output analysis was used to determine the economic impacts of the recreational boating on Donner Lake. Donner Lake is used as an example in this economic impact assessment because quantifiable data in terms of boating survey are available. The analysis used here can be used with any other body of water for which such data exist-.

Estimates of total trip expenditures were developed from data on the annual boating expenditures available from a study done for the California Department of Boating and Waterways (California Department of Boating and Waterways, 1997) in combination with the 2000-boating season survey conducted for this guidance manual. Recreational boating in Donner Lake occurs over a shortened season than elsewhere in the State. As such, trip expenditures would need to be adjusted to reflect the difference. In the absence of representative data showing differences in seasonal length, an assumption was made to use the boating trip expenditures from the California Department of Boating and Waterways as reported in the study.

Trip-related expenditures (e.g., food, lodging, transportation, and activities) are typically divided into three groups: those made at or near home, those made en route to and from the recreation site, and those made at or near the recreation site. Only expenditures made by nonresidents are relevant for determining economic impacts. These expenditures would include all expenditures made at or near the site as well as a portion of the expenditures made en route. For example, the following proportions were applied to individual expenditure categories to arrive at the proportion of those expenditures that stay within the local economy. For restaurants, it was assumed that 90 percent of expenditures would remain in the county. Thirty percent of expenditures on grocery items and 50 percent of all other expenditure categories were assumed to remain within the county. The preceding proportions represent the local contribution and are partially based on estimates developed by Clawson and Knetch (1966) for economies of rural recreational sites near federal reserves. In the absence of better estimates, the preceding proportions were used.

For this analysis, the following additional assumptions were made:

- The region of influence for the economic analysis is Nevada County.
- 40 percent of the visitors to Donner Lake are nonresidents of Nevada County.
- A partial ban on recreational boating on Donner Lake would involve a 35 percent reduction in the number of boats.
- The base year of analysis is 1997, but the impacts were adjusted to reflect year 2000 price levels.

DATA

Two data sources were used to derive the total trip expenditures by expenditure category:

- Total number of boats on Donner Lake from the 2000 boating survey.
- Annual boating trip expenditure data from the California Department of Boating and Waterways Economic Impact of Boating in California study.

The 2000 boating survey indicated there were 8,825 boats on Donner Lake during the season. Table J.1 shows the annual boating trip expenditures by expenditure category as well as the total annual boating expenditures that stay in Nevada County for each of these expenditure categories.

Table J.1

Annual boating trip expenditures, 1997 $

Expenditure Category	Annual Per Boat Trip Expenditures	Proportion of Expenditure assumed Local	Total Annual Boating Trip Expenditures ($)[a]
Food	$1,165	30%	$1,299,308
Restaurant	$401	90%	$1,341,688

(continued)

Table J.1 (continued)

Expenditure Category	Annual Per Boat Trip Expenditures	Proportion of Expenditure assumed Local	Total Annual Boating Trip Expenditures ($)[a]
Fishing Equipment	$101	50%	$187,740
Equipment Rental	$12	50%	$22,306
Lodging	$448	50%	$832,746
Fuel	$536	50%	$996,322
Boat Launching[b]		100%	$7,024
Equipment & Supplies	$87	50%	$161,270
Repair & Maintenance	$357	50%	$663,688
Insurance, fees & taxes	$143	50%	$266,553
TOTAL	$3,250		$5,778,645

Source: Data from California Department of Boating and Waterways, 1997.
a Total expenditures are derived by multiplying the second and third columns of the table by the total number of boats (8,825) and adjusting those expenditures by the percentage of boats that are from outside Nevada County (i.e., 40%).
b Boat Launching (including vehicle parking fees) fees are those reported by the Truckee Recreation and Parks Department for the 2000 boating season.

The estimated total trip expenditures were analyzed to estimate the baseline economic worth of the recreational boating industry to Nevada County. An IMPLAN regional input-output model was constructed for the county. The expenditure data in the Table J.1 form the input into the model.

Estimating the Baseline Economic Worth of Recreational Boating on Donner Lake

Recreational boating on Donner Lake contributes about 145 jobs in direct employment and about $2.4 million in direct personal income to the Nevada County economy. In addition to the direct economic contributions of visitor expenditures, recreational boating on Donner Lake also contributes to the economic well-being of Nevada County through secondary economic impacts (indirect and induced impacts). Visitor expenditures result in 20 indirect and 22 induced employment.

The visitor expenditures also generate $0.5 million each in indirect and induced personal income to the region. Table J.2 shows the estimates of direct, indirect and induced employment and personal income generated by recreational boating on the Nevada County economy.

Table J.2

Economic impact estimates of recreational boating, Nevada County

	Estimate of Economic Impact
Employment	
Direct	145 full-time equivalents
Indirect	20 full-time equivalents
Induced	22 full-time equivalents
Total Employment	188 full-time equivalents
Personal Income	
Direct	$2.4 million
Indirect	$0.5 million
Induced	$0.5 million
Total Income	$3.4 million

Source: CH2MHILL 2002.
Income estimates are in 2000 $.
Numbers may not add up due to rounding.

Recreational boating on Donner Lake is an important economic activity in Nevada County. Total employment due to boating represent about 0.7 percent of the total regional industry employment of 27,970 (State of California Employment Development Department 2002). Total personal income due to recreational boating, on the other hand, represent about 0.1 percent of total regional personal income (here derived as per capita income of $26,341 multiplied by the Census 2000 population estimate for Nevada County of 92,033 (State of California Department of Finance 2002)). Thus, recreational boating on Donner Lake has a beneficial impact on regional employment and personal income in Nevada County.

Estimating the Economic Impact of a Partial Ban on Recreational Boating on Donner Lake

A partial ban on recreational boating on Donner Lake would have the effect of reducing the number of boats by 35 percent (the percentage of boats with two-stroke engines from the 2000 boating survey). The expenditure data in Table J.1, adjusted to reflect the 35 percent reduction in recreational boats, were used as input into the Nevada County Input/Output IMPLAN model. Table J.3 shows the estimates of direct, indirect and induced employment and personal income associated with the 35 percent reduction in recreational boating on Donner Lake. All the estimates in Table J.3 are 35 percent of the estimates in Table J.2.

Table J.3

Economic impact estimates of partial banning of recreational boating, Nevada County

	Estimate of Economic Impact
Employment	
Direct	51 full-time equivalent positions
Indirect	8 full-time equivalent positions
Induced	7 full-time equivalent positions
Total Employment	66 full-time equivalent positions
Personal Income	
Direct	$0.8 million

(continued)

Table J.3 (continued)

	Estimate of Economic Impact
Indirect	$0.2 million
Induced	$0.2 million
Total Income	$1.1 million

Source: CH2MHILL 2002.
Income estimates are in 2000 $.
Numbers may not add up due to rounding.

The loss of all or some of the recreational opportunities (resulting from a total or partial ban on boating) would have the effect of significantly reducing visitor expenditures. Any reduction in visitor expenditures would result in a negative economic impact (at least in the short to medium term or before resources are reallocated elsewhere) in areas with surface water bodies that are currently also used for recreational purposes such as Donner Lake.

APPENDIX K
COST SUMMARY FOR TREATMENT TECHNOLOGIES

Table K.1

Cost summary for packed tower aeration

Flow	Influent Concentration (ppb)	Effluent Concentration (ppb)	Removal Efficiency (Percent)	Cost ($/1,000 gallons)
60	20	0.5	97.5	1.75
		5	75	1.66
	200	0.5	99.75	1.82
		5	97.5	1.75
		20	90	1.66
	2,000	0.5	99.98	ND
		5	99.75	1.82
		20	99	1.79
600	20	0.5	97.5	0.34
		5	75	0.3
	200	0.5	99.75	0.37
		5	97.5	0.34
		20	90	0.32
	2,000	0.5	99.98	ND
		5	99.75	0.37
		20	99	0.36

(continued)

Table K.1 (continued)

Flow	Influent Concentration (ppb)	Effluent Concentration (ppb)	Removal Efficiency (Percent)	Cost ($/1,000 gallons)
6,000	20	0.5	97.5	0.16
		5	75	0.13
	200	0.5	99.75	0.17
		5	97.5	0.16
		20	90	0.15
	2,000	5	99.75	0.18
		20	99	0.17

Capital expenses include equipment, piping, valves and electrical components (30 percent), site work (10 percent), contractor O&P (15 percent) and contingency (20 percent).

O&M costs include: power costs at $0.08/kWhr, labor costs estimated at $80/hr and analytical costs estimated at $200 per sample.

Table K.2

Cost summary for PTA with off-gas carbon adsorption

Flow (gpm)	Influent MTBE Concentration (ppmv)	Cost ($/1,000 gallons)
60	0.5	0.54
	5	1.86
600	0.5	0.24
	5	1.56
6,000	0.5	0.23
	5	1.55

O&M costs include: power costs (at $0.08 / kWhr), fuel costs (at $5.00 / Mbtu), analytical costs estimated at $200 per sample, GAC changeout at $1.5/lb, use rate of 0.05 lb/1000 cubic feet air at 5 ppmv MTBE and 0.006 lb/ 1000 cubic feet at 0.5 ppmv MTBE.

Table K.3

Cost summary for advanced oxidation with ultraviolet light and hydrogen peroxide

Flow	Influent Concentration (ppb)	Effluent Concentration (ppb)	Removal Efficiency (Percent)	Cost ($/1,000 gallons)
60	20	0.5	97.5	2.50
		5	75	2.18
	200	0.5	99.75	2.72
		5	97.5	2.50
		20	90	2.32
	2,000	0.5	99.98	4.11
		5	99.75	3.47
		20	99	3.07
600	20	0.5	97.5	0.91
		5	75	0.57
	200	0.5	99.75	1.27
		5	97.5	0.96
		20	90	0.71
	2,000	0.5	99.98	2.08
		5	99.75	1.75
		20	99	1.52
6,000	20	0.5	97.5	0.52
		5	75	0.32
	200	0.5	99.75	0.74
		5	97.5	0.6
		20	90	0.42

(continued)

Table K.3 (continued)

Flow	Influent Concentration (ppb)	Effluent Concentration (ppb)	Removal Efficiency (Percent)	Cost ($/1,000 gallons)
	2,000	0.5	99.98	1.59
		5	99.75	1.24
		20	99	0.65

Table K.4

Cost summary for advanced oxidation with ozone and hydrogen peroxide

Flow	Influent Concentration (ppb)	Effluent Concentration (ppb)	Removal Efficiency (Percent)	Cost ($/1,000 gallons)
60	20	0.5	97.5	2.68
		5	75	2.63
	200	0.5	99.75	2.98
		5	97.5	2.68
		20	90	2.65
	2,000	0.5	99.98	3.62
		5	99.75	3.31
		20	99	3.29
600	20	0.5	97.5	0.9
		5	75	0.82
	200	0.5	99.75	0.95
		5	97.5	0.90
		20	90	0.84

(continued)

Table K.4 (continued)

Flow	Influent Concentration (ppb)	Effluent Concentration (ppb)	Removal Efficiency (Percent)	Cost ($/1,000 gallons)
	2,000	0.5	99.98	1.19
		5	99.75	1.13
		20	99	1.07
6,000	20	0.5	97.5	0.43
		5	75	0.35
	200	0.5	99.75	0.48
		5	97.5	0.43
		20	90	0.37
	2,000	0.5	99.98	0.68
		5	99.75	0.59
		20	99	0.56

Table K.5

Cost summary for granular activated carbon

Flow	Influent Concentration (ppb)	Cost ($/1,000 gallons)
60	20	2.3
	200	2.92
	2,000	4.43
600	20	0.77
	200	1.15
	2,000	2.37
6,000	20	0.5
	200	0.97

(continued)

Table K.5 (continued)

Flow	Influent Concentration (ppb)	Cost ($/1,000 gallons)
	2000	2.22

Cost estimates have been developed for in-series operation

Effluent contains no detectable MTBE (< 0.5 μg/l)

REFERENCES

Allen, B.C. and J.E. Reuter. 1999. Changes in MTBE and BTEX Concentrations in Lake Tahoe, CA-NV Following Implementation of a ban on Selected Two-stroke Marine Engines. John Muir Institute for the Environment, University of California, Davis.

Allen, B.C., J.E. Reuter, C.R. Goldman, M.F. Fiore, and G.C. Miller. 1998. Lake Tahoe Motorized Watercraft - an Integration of Water Quality, Watercraft use and Ecotoxicology Issues. John Muir Institute for the Environment, University of California, Davis.

Anderson, M. 1997. Predicted MTBE concentrations in the Eastside Reservoir. Report to the Los Angeles Municipal Water District.

Atkins, P.W. *Physical Chemistry*. 1986. New York: W.H. Freeman and Company.

Arizona Department of Environmental Quality (ADEQ). 2001 http://www.adeq.state.az.us/. (August 22, 2001).

Baehr, A.L. and Reilly, T.J. 2001. Water Quality and Occurrence of Methyl-tert Butyl Ether (MTBE) and Other Fuel Related Compounds in Lakes and Groundwater at Lakeside Communities in Sussex and Morris Counties, New Jersey, 1998-1999. U.S. Geological Survey Water-Resources Investigations Report, in press.

Baehr, A.L., P.E. Stackelberg, and R.J. Baker. 1999. Evaluation of the Atmosphere as a Source of Volatile Organic Compounds in Shallow Groundwater. *Water Resources Research* 35 (1): 127-136.

Beckel, A. 2001. Email Communication with A. Rossi, CH2M HILL. January 30.

Bender, D.A., W.E. Asher, and J.S. Zogorski. 2002. Theoretical basis, user instructions, and verification of a deterministic model to estimate volatile organic compound concentrations in reservoirs and lakes: Draft, U.S. Geological Survey Open-File Report.

Berger, Robert. 2001. Personal communication with Radha Gapolan, CH2M HILL. August 22, 2001

Boughton, C.J. and M.S. Lico. 1998. Volatile Organic Compounds in Lake Tahoe, Nevada and California, July-September 1997. USGS Fact Sheet FS-055-98, June.

California Bowen Bill. 1998. California Bowen Bill AB2439 Press Coverage. http://www.rbbi.com/desks/regs/ca2439/capress.htm

California Department of Boating and Waterways, 1997. *The Economic Impact of Boating in California.* Prepared by the Public Research Institute, San Francisco State University, San Francisco, CA.

California Department of Health Services (CA DHS). 2001. http://www.dhs.ca.gov/ps/ddwem/chemicals/MTBE/mtbeindex.htm (March 10, 2001).

California MTBE Research Partnership. 2000. Treatment Technologies for Removal of Methyl Tertiary Butyl Ether (MTBE) From Drinking Water: Air Stripping; Advanced Oxidation Processes; Granular Activated Carbon; Synthetic Resin Sorbents. Edited by Gina Melin, National Water Research Institute Second Edition, February. Published by the Center for Groundwater Restoration and Protection, National Water Research Institute, California.

CARB (California Air Resources Board). 2001. Outboard Engine and Personal Watercraft Emissions to Air and Water: A Laboratory Study. Mobile Source Control Division, Monitoring and Laboratory Division, California Air Resources Bd., El Monte CA.

Chapra, Steven C. 1997. *Surface water-quality modeling*. New York : McGraw-Hill.

Clawson, M. and J. L. Knetch. 1966. *Economics of Outdoor Recreation.* Baltimore MD: John Hopkins Press.

Dale, M.S., B. Koch, R.F. Losee, E.W. Crofts, and M.K. Davis. 2000. MTBE in Southern California Water. *Journal AWWA*, 92(8): 42-51.

Dong, A. E. *Limnological Data for Donner Lake, California May 1973 Through December 1973*; U.S. Geological Survey: Menlo Park, CA, 1975; Open-File Report 6226-01.

Drinking Water Advisory: Consumer Acceptability Advice and Health Effects Analysis on Methyl Tertiary-Butyl Ether (MTBE). 1997. Fact sheet. and Advisory. EPA-822-F-97-009. ODW 4304. Health and Ecological Criteria Division, Office of Science and Technology, Office of Water, U.S. Environmental Protection Agency. Washington, D.C.: U.S. EPA. December.

European Fuel Oxygenates Association, 2000. *MTBE Resource Guide, Appendix 24: MTBE Treatability 2000.* http://www.efoa.org.

Fiore, M.F., C. Hoonhout, V. Hebert, J. Herz, S. Sotomey, and G.C. Miller. 1998. Interim Report on the Lake Tahoe Motorized Watercraft Study. Department of Environmental and Resource Sciences, University of Nevada, Reno.

Fischer, H. B., E.J. List, R. C. Y. Koh, J. Imberger, and N. H. Brooks. 1979. Mixing in Inland and Coastal Waters. New York: Academic Press.

Frysinger, G.S., R.B. Gaines, and E.B. Ledford. 1999. Quantitative determination of BTEX and total aromatic compounds in gasoline by comprehensive two-dimensional gas chromatography (GC x GC). *HRC-Journal of High-Resolution Chromatography.* April, v. 22(4): 195-200.

Gabele, P.A. and S.M. Pyle. 2000. Emissions from Two Outboard Engines Operating on Reformulated Gasoline Containing MTBE. *Environ. Sci. Technol.* 34: 368-372.

Hayduk, W. and H. Laudie. 1974. Prediction of diffusion coefficients for nonelectrolytes in dilute aqueous solutions. *Amer. Inst. Chem. Engrg. Journal*, 20: 611-615.

Hagler Bailly, Inc. 1998. Revised Boating Fuel Use Estimate Tables. Technical Memorandum from Bob Baumgartner to Gordon Barrett, Tahoe Regional Planning Agency. October 27.

Hamann, Carl L. Jr., McEwen, J.B., and Myers, A.G. 1990. "Guide to Selection of Water Treatment Processes." Water Quality and Treatment. Ed. Frederick W. Pontius. McGraw-Hill Inc, New York.

Hamilton, D. P. and S.G. Schladow. 1996. Prediction of Water Quality in Lakes and Reservoirs. Part I - Model Description. *Ecol Modeling* 96: 91-110.

Hanson, J.R., Ackerman, C.E., Scow, K.M. (1999). "Biodegradation of Methyl tert-Butyl Ether by a Bacterial Pure Culture." *Appl. Environ. Microbiol.*, pp. 4788-4792.

Heermann, S.E. and S.E. Powers. 1998. Modeling the partitioning of BTEX in water-reformulated gasoline systems containing ethanol. *Journal of Contaminant Hydrology*. Nov. 15, v. 34(4):315-341.

Hitzig, R., Kostecki, P. and Leonard, D. 1998. Study reports LUST programs are feeling effects of MTBE releases. Soil & Groundwater Cleanup August/September Issue, pp. 15-19.

Hocking, G.C. and J.C. Patterson. 1991. Quasi-two-dimensional reservoir simulation model. *J. Environ Eng*, ASCE, 117(5): 595-613.

Howard, P.H., R.S. Boethling, W.F. Jarvis, W.M. Meylan, and E. M. Michalenko. 1991. *Environmental degradation rates*. Chelsea, Mich: Lewis Publishers (2nd printing).

Idaho Department of Health and Welfare (IDHW). 2001. http://www2.state.id.us/dhw/. (August 22, 2001).

Imberger, J. and J.C. Patterson. 1981. A Dynamic Reservoir Simulation Model - DYRESM:5, in *Transport Models for Inland and Coastal Waters*. edited by H.B. Fischer. New York: Academic Press.

Imberger, J., J. Patterson, B. Hebbert, and I. Loh. 1978. Dynamics of Reservoirs of Medium Size. *J. Hydraul. Div.*, ASCE 104: 725-743.

Juttner, F., D. Backhaus, U. Matthias, U. Essers, R. Greiner, and B. Mahr. 1995. Emissions of two- and four-stroke outboard engines - I. Quantification of gases and VOC. *Water Res.*, 29: 1976-1982.

Kalman, O. and R. Lund. 1998. MTBE: Evaluation of Management Options for Water Supply and Ecosystem Impacts. Draft Report. Dept. of Civil and Environmental Engineering, University of California, Davis.

Kaufmann, N., Truckee-Donner Utility. 2001. Telephone Conversation with Radha Gopalan, CH2M HILL. October 1.

Keller, A, J. Friones, C. Koshland, J. Reuter, M. Suffet, and J. Last. 1998. Health & Environmental Assessment of MTBE. Report to the Governor and Legislature of the State of California as Sponsored by SB 521. Volume I - Summary & Recommendations. University of California Toxic, Substances Research & Teaching Program, Davis, CA.

Lewis, W. K. and W. G. Whitman. 1924. Principles of gas absorption. *Ind. Engrg. Chem.*, 16, 1215-1220.

Maine Department of Environmental Protection (ME DEP). http://janus.state.me.us/dep/home.htm (March 10, 2001).

Maryland Department of the Environment (MA DE). 2001. http://www.mde.state.md.us/. (August 22, 2001).

Massachusetts Department of Environmental Protection (MA DEP). 2001. http://www.state.ma.us/dep/dephome.htm. (August 22, 2001).

McCord, S.A. and S.G. Schladow. 1998. Transport and Fate Modeling of MTBE in Lakes and Reservoirs. In: Health & Environmental Assessment of MTBE. Report to the Governor and Legislature of the State of California as Sponsored by SB 521. Volume IV - Ground & Surface Water. University of California Toxic, Substances Research & Teaching Program, Davis, CA.

Malcom-Pirnie, Inc. 1998. Modeling the volatilization of methyl tertiary-butyl ether (MTBE) from surface impoundments. Technical Memorandum to the Oxygenated Fuels Association, Inc., April 3.

Michigan Department of Environmental Quality (MI DEQ). 2001.http://www.michigan.gov/deq. (August 22, 2001).

Mower, Barry. 2001. Personal Communication with A. Rossi, CH2M HILL. January 15.

MTBE Work Group and Division of Science, Research and Technology Regulations (http://www.state.nj.us/dep/dsr/mtbe/fate_and_transport.htm

New Hampshire Department of Environmental Services (NHDES). http://www.des.state.nh.us/water_intro.htm (March 10, 2001).

New Hampshire Department of Health and Human Services (NH DHHS). http://www.dhhs.state.nh.us (March 10, 2001).

New Jersey Department of Environment. 2000, MTBE IN NEW JERSEY'S ENVIRONMENT.
New Jersey Department of Environmental Protection (NJ DEP). 2001. http://www.state.nj.us/dep/dsr/mtbe/regulation-guidance.htm (March 6, 2001).

New York State Department of Environmental Conservation (NY DEC). 2001. http://www.dec.state.ny.us/website/dow/index.html (March 6, 2001).

North Carolina Department of Environment and Natural Resources (NC DENR). 2001. http://www.enr.state.nc.us/. (August 22, 2001).

Pankow, J.F., N.R. Thomson, R.L. Johnson, A.L. Baehr, and J.S. Zogorksi. 1997.The urban atmosphere as a non-point source for the transport of MTBE and other volatile organic compounds (VOCs) to shallow groundwater. *Environmental Science & Technology*, Oct., v31(N10):2821-2828.

Pankow, J. F., R. E. Rathbun, and J. S. Zogorski. 1996. Calculated volatilization rates of fuel oxygenate compounds and other gasoline-related compounds from rivers and streams. *Chemosphere*, 33(5): 921-937.

Patterson, J. C., P.F. Hamblin, and J. Imberger. 1984. Classification and dynamics simulation of the vertical density structure of lakes. *Limnol. Oceanogr.*, 29: 845 - 861.

Pennsylvania Department of Environmental Protection (PA DEP). 2001. http://www.dep.state.pa.us/. (August 22, 2001).

Rathbun, R.E. 1998. Transport, Behavior, and Fate of Volatile Organic Compounds in Streams. U.S. Geological Survey.

Reuter, J.E., B.C. Allen, R.C. Richards, J.F. Pankow, C.R. Goldman, R.L. Scholl, and J.S. Seyfried. 1998. Concentrations, Sources, and Fate of the Gasoline Oxygenate Methyl tert-butyl ether (MTBE) in a Multiple-use Lake. *Environ. Sci. Technol.* 32: 3666-3672.

Rhode Island Department of Health (RI DH). 2001. http://www.health.state.ri.us/. (August 22, 2001).

Robbins, G. A., S. Wang, and J. D. Stuart. 1993. Using the static headspace method to determine Henry's Law constants. *Analytical Chemistry*, 85: 3113-3118.

Salanitro, J.P., et. al., 2001. In Situ Bioremediation of MTBE Using Biobarriers Of Single Or Mixed Cultures in Proceedings of 6th Int'l Symp. On In Situ and On-Site Bioremediation, Battelle Press, San Diego, In Press.

Schladow, S.G. 1993. Destratification of Avon Reservoir. Report to Sydney Water Board. Centre for Water Research Report WP 781 GS.

Schladow, S.G. and D.P. Hamilton. 1996. Prediction of Water Quality in Lakes and Reservoirs. Part II - Application to Prospect Reservoir. *Ecol Modeling* 96: 111-12.

Schwarzenbach, R. P., P. M. Gschwend, and D. M. Imboden. 1993. *Environmental Organic Chemistry.* New York: Wiley-Interscience

State of California Department of Boating and Waterway. 1997. The Economic Impact of Boating in California. Prepared by the Public Research Institute, San Francisco State University and Planning and Applied Economics, Berkeley, California.

State of California Department of Finance. 2002. *Demographic Information – 2000 Census.* http://www.dof.ca.gov/HTML/DEMOGRAP/2000Cover1.htm

State of California Department of Finance. 2002. *Financial and Economic Data – County Profiles.* http://www.dof.ca.gov/HTML/FS_DATA/profiles/pf_home.htm

State of California Employment Development Department. 2002. *Labor Market Information – Employment by industry data.* http://www.calmis.ca.gov/htmlfile/subject/indtable.htm

Talbot, Roger. 2000. "NH Starts Toughest Standard on MTBE." *The Union Leader and New Hampshire Sunday News.* May 14. www.theunionleader.com/pages/water13.htm (March 6, 2000).

Tikkanen, M.; 1998. Memo on the Cost of Monitoring for MTBE on East Bay Municipal Utility District's Drinking Water Reservoirs, East Bay Municipal Utility District, Oakland.

U.S. Environmental Protection Agency (USEPA). 2000. http://www.epa.gov/ogwdw000/mtbe.html. Updated on November 30, 2000. Accessed on January 18, 2001.

USEPA 1986. Nineteenth report of the interagency committee to the administration; receipt and request for comments regarding priority list of chemicals. Federal Register, 51(210): 41,417-41,432.

USEPA. 1997.Drinking Water Advisory – Consumer Acceptability Advice and Health Effects Analysis on Methyl Tertiary Butyl Ether: Washington, D.C., Office of Water, EPA 822-F97-009.

USEPA. 2001. Methyl Tertiary Butyl Ether. http://www.epa.gov/mtbe/faq.htm

USEPA. 2001b. Safe Drinking Water Hotline, 800/426-4791. Telephone Conversation with D. Sokolove, CH2M HILL. January 19.

Vermont Department of Environmental Conservation (VT DEC). 2001. http://www.anr.state.vt.us/dec/dec.htm. (August 22, 2001).

Vineyard, G.L. 1994. Results from Monitoring Program in Donner Lake 1991-1993. Department of Biology, University of Nevada, Reno. Unpublished report.

Wanninkhof, R., J. R. Ledwell, and J. Crusius. 1991. Gas transfer velocities on lakes measured with sulfur hexafluoride. In: *Air-Water Mass Transfer*. Edited by S.C. Wilhelms and J.S. Gulliver. New York: American Society of Civil Engineers.

Weast, R.C. 1983. CRC *Handbook of Chemistry and Physics*, 63rd Edition. Boca Raton, Fla.: CRC Press Inc.

Weinstock, J. 1981. Vertical turbulence diffusivity for weak or strong stable stratification. *J. Geophys.* Res., 86(C10): 9925-9928.

Wisconsin Department of Natural Resources (WI DNR). 2001. http://www.dnr.state.wi.us. (August 22, 2001).

Wyoming Department of Environmental Quality (WY DEQ). 2001. http://deq.state.wy.us/. (August 22, 2001).

Zogorski, J.S., A.L. Baehr, B.J. Bauman, D.L. Conrad, R.T. Drew, N.E. Korte, W.W. Lapham, A. Morduchowitz, and J.F. Pankow. 1996. Fuel Oxygenates and Water Quality: Current Understanding of Sources, Occurrence in Natural Waters, Environmental Behavior, Fate and Significance. Office of Science and Technology Policy Report; Executive Office of the President, Washington, D.C.

ABBREVIATIONS

°C	degrees Celsius
µS	micro Siemens
µg/L	micrograms per liter
AOP	advanced oxidation process
atm	atmosphere
AWWA	American Water Works Association
AwwaRF	Awwa Research Foundation
BHRA	Bureau of Health Risk Assessment
BTEX	benzene, toluene, ethylbenzene, and xylene
CA DHS	California Department of Health Services
CA DTSC	California Department of Toxic Substances Control
CalEPA	California Environmental Protection Agency
CARB	California Air Resources Board
CCL	Contaminant Candidate List
CCWD	Contra Costa Water District
cm/d	centimeters per day
CO	carbon monoxide
CVO ASL	CH2M Hill's Corvallis Applied Sciences Laboratory
DEQ	Arizona Department of Environmental Quality
DHW	Idaho Department of Health Welfare
DIPE	disopropyl ether
DLM	Dynamic Lake Model
DME	dimethyl ether

EBMUD	East Bay Municipal Utility District
ETBE	ethyl tertiary butyl ether
g/L	gallons per liter
GAC	granular activated carbon
GC-MSD	gas chromatography and mass spectrometry detection
gpm	gallons per minute
Hg	Mercury
Hp	horsepower
kg	kilograms
kj/m^2•d	kilojoules per square meter day
km	kilometer
km^2	square kilometer
Kow	partition coefficient
kPa	kiloPascal
LIMS	laboratory information management system
LOD	limit of detection
m	meter
m/sec	meter per second
m^2/s	meter squared per second
m^3/day	cubic meter per day
mbar	millibar
MCLs	maximum contaminant levels
MCLGs	maximum contaminant level goals
MA DE	Maryland Department of the Environment
ME DEP	Maine Department of Environmental Protection
mg	milligram

mg/L	milligrams per liter
MI DEQ	Michigan Department of Environmental Quality
mm	millimeter
MSD	mass spectrometer detector
MTBE	methyl tertiary butyl ether
MWDSC	Metropolitan Water District of Southern California
NC DENR	North Carolina Department of Environment and Natural Resources
NDMA	N-nitroso dimethyl amine
NH DHHS	New Hampshire Department of Health and Human Services
NJ DEP	New Jersey Department of Environmental Protection
NOM	natural organic matter
NPDWR	National Primary Drinking Water Regulation
NSDWR	National Secondary Drinking Water Regulations
NY DEC	New York State Department of Environmental Conservation
O&M	Operations and Maintenance
OFA	Oxygenated Fuels Association
OMC	Outboard Marine Corporation
PA DEP	Pennsylvania Department of Environmental Protection
PAH	polynuclear aromatic hydrocarbon
Partnership	California MTBE Research Partnership
PCE	perchloroethylene
ppb	parts per billion
PTA	packed tower aeration
PWC	Personal Watercraft
QA/QC	quality assurance/quality control

RBR	Richard Brancker Research, Ltd.
RFG	reformulated gasoline
RI DH	Rhode Island Department of Health
SB 521	California Senate Bill No. 521 (1997)
SDWA	Safe Drinking Water Act
SMCL	secondary maximum containment level
SOC	synthetic organic chemicals
SWRCB	California State Water Resources Control Board
TAEE	tertiary amyl ethyl ether
TAME	tertiary amyl methyl ether
TBA	tertiary butyl alcohol
TBF	tertiary-butyl formate
TCE	trichloroethylene
TRG	Tahoe Research Group
UCD	University of California at Davis
UCMR	Unregulated Contaminant Monitoring Rule
UNR	University of Nevada at Reno
USEPA	United States Environmental Protection Agency
USGS	United States Geological Survey
UV	ultraviolet
VOC	volatile organic compound
VT DEC	Vermont Department of Environmental Conservation
WI DNR	Wisconsin Department of Natural Resources
WSPA	Western States Petroleum Association
w/w	weight by weight
WY DEQ	Wyoming Department of Environmental Quality

IWA Publishing's authorised EU representative for General Product Safety Regulations is Diane D'Arras, 15 rue Duret, 75116 Paris, France, e-mail: safety@iwap.co.uk.

Printed and bound by CPI Group (UK) Ltd, Croydon, CR0 4YY
18/09/2026
02218938-0006